Modern Directional Statistics

CHAPMAN & HALL/CRC

Interdisciplinary Statistics Series

Series editors: N. Keiding, B.J.T. Morgan, C.K. Wikle, P. van der Heijden

Published titles

AGE-PERIOD-COHORT ANALYSIS: NEW MODELS, METHODS, AND EMPIRICAL APPLICATIONS Y. Yang and K. C. Land

ANALYSIS OF CAPTURE-RECAPTURE DATA R. S. McCrea and B. J.T. Morgan

AN INVARIANT APPROACH TO STATISTICAL ANALYSIS OF SHAPES S. Lele and J. Richtsmeier

ASTROSTATISTICS G. Babu and E. Feigelson

BAYESIAN ANALYSIS FOR POPULATION ECOLOGY R. King, B. J.T. Morgan, O. Gimenez, and S. P. Brooks

BAYESIAN DISEASE MAPPING: HIERARCHICAL MODELING IN SPATIAL EPIDEMIOLOGY, SECOND EDITION A. B. Lawson

BIOEQUIVALENCE AND STATISTICS IN CLINICAL PHARMACOLOGY S. Patterson and B. Jones

CLINICAL TRIALS IN ONCOLOGY, THIRD EDITION S. Green, J. Benedetti, A. Smith, and J. Crowley

CLUSTER RANDOMISED TRIALS R.J. Hayes and L.H. Moulton

CORRESPONDENCE ANALYSIS IN PRACTICE, THIRD EDITION M. Greenacre

THE DATA BOOK: COLLECTION AND MANAGEMENT OF RESEARCH DATA M. Zozus

DESIGN AND ANALYSIS OF QUALITY OF LIFE STUDIES IN CLINICAL TRIALS, SECOND EDITION D.L. Fairclough

DYNAMICAL SEARCH L. Pronzato, H. Wynn, and A. Zhigljavsky

FLEXIBLE IMPUTATION OF MISSING DATA S. van Buuren

GENERALIZED LATENT VARIABLE MODELING: MULTILEVEL, LONGITUDINAL, AND STRUCTURAL EQUATION MODELS A. Skrondal and S. Rabe-Hesketh

GRAPHICAL ANALYSIS OF MULTI-RESPONSE DATA K. Basford and J. Tukey

INTRODUCTION TO COMPUTATIONAL BIOLOGY: MAPS, SEQUENCES, AND GENOMES M. Waterman

MARKOV CHAIN MONTE CARLO IN PRACTICE W. Gilks, S. Richardson, and D. Spiegelhalter

MEASUREMENT ERROR ANDMISCLASSIFICATION IN STATISTICS AND EPIDEMIOLOGY: IMPACTS AND BAYESIAN ADJUSTMENTS P. Gustafson

Published titles

MEASUREMENT ERROR: MODELS, METHODS, AND APPLICATIONS
J. P. Buonaccorsi

MEASUREMENT ERROR: MODELS, METHODS, AND APPLICATIONS
J. P. Buonaccorsi

MENDELIAN RANDOMIZATION: METHODS FOR USING GENETIC VARIANTS IN CAUSAL ESTIMATION S.Burgess and S.G.Thompson

META-ANALYSIS OF BINARY DATA USINGPROFILE LIKELIHOOD D. Böhning, R. Kuhnert, and S. Rattanasiri

MISSING DATA ANALYSIS IN PRACTICE T. Raghunathan

MODERN DIRECTIONAL STATISTICS C. Ley and T. Verdebout

POWER ANALYSIS OF TRIALS WITH MULTILEVEL DATA M. Moerbeek and S. Teerenstra

SPATIAL POINT PATTERNS: METHODOLOGY AND APPLICATIONS WITH R
A. Baddeley, E Rubak, and R. Turner

STATISTICAL ANALYSIS OF GENE EXPRESSION MICROARRAY DATA T. Speed

STATISTICAL ANALYSIS OF QUESTIONNAIRES: A UNIFIED APPROACH BASED ON R AND STATA F. Bartolucci, S. Bacci, and M. Gnaldi

STATISTICAL AND COMPUTATIONAL PHARMACOGENOMICS R. Wu and M. Lin

STATISTICS IN MUSICOLOGY J. Beran

STATISTICS OF MEDICAL IMAGING T. Lei

STATISTICAL CONCEPTS AND APPLICATIONS IN CLINICAL MEDICINE
J. Aitchison, J.W. Kay, and I.J. Lauder

STATISTICAL AND PROBABILISTIC METHODS IN ACTUARIAL SCIENCE
P.J. Boland

STATISTICAL DETECTION AND SURVEILLANCE OF GEOGRAPHIC CLUSTERS
P. Rogerson and I. Yamada

STATISTICS FOR ENVIRONMENTAL BIOLOGY AND TOXICOLOGY A. Bailer and W. Piegorsch

STATISTICS FOR FISSION TRACK ANALYSIS R.F. Galbraith

VISUALIZING DATA PATTERNS WITH MICROMAPS D.B. Carr and L.W. Pickle

Chapman & Hall/CRC
Interdisciplinary Statistics Series

Modern Directional Statistics

Christophe Ley • Thomas Verdebout

CRC Press
Taylor & Francis Group
Boca Raton London New York

CRC Press is an imprint of the
Taylor & Francis Group, an informa business
A CHAPMAN & HALL BOOK

CRC Press
Taylor & Francis Group
6000 Broken Sound Parkway NW, Suite 300
Boca Raton, FL 33487-2742

Printed on acid-free paper by Ashford Colour Press Ltd.
Version Date: 20170522

International Standard Book Number-13: 978-1-4987-0664-3 (Hardback)

Contents

Preface

Aim of the book

The field of directional statistics has received a lot of attention over the past two decades. This is mainly due to new demands from domains such as bioinformatics, machine learning and cosmology, but also to the availability of massive data requiring adapted statistical techniques, and to technological advances. Our goal is to provide a thorough overview of the developments that have taken place since the beginning of the 21st century, hereby building upon earlier works like the seminal and inspiring monograph by Mardia & Jupp (2000). We shall focus here solely on theoretical developments; the description of modern datasets and their analysis will be provided in the companion book *Applied Directional Statistics: Modern Methods and Case Studies* for which the authors are acting as editors.

Complementarity with existing literature

The present book has a well-defined target: reference the methodological improvements on directional statistics that have appeared after the cornerstone reference books by Mardia & Jupp (2000) and Jammalamadaka & SenGupta (2001). This has so far been only partially done by Pewsey et al. (2013) but specifically for the circular setting and with more focus on applications. The present book fills this gap and will therefore be of interest for researchers and practitioners dealing with directional data and aiming for a recent methodological flavour.

Acknowledgments

There are a number of people we wish to thank for their contributions to our book. We thank our editor, John Kimmel, for insightful exchanges and for the initial proposal to write this book. We are grateful to Arthur Pewsey for a very careful reading of the entire book, to Toshihiro Abe, Rosa Crujeiras, Yves Dominicy,

Oliver Dukes, Eduardo García-Portugués, Thomas Hamelryck, Shogo Kato, Irene Klugkist, Davy Paindaveine, Thanh Mai Pham Ngoc and Yvik Swan for their helpful comments and suggestions, to José Ameireijas-Alonso for his help with some of the graphics, and to Kim Schalbar for checking the bibliography.

Our very special thanks go to our families, to whom we dedicate this book.

Introduction

1.1 Overview

1.1.1 A brief introduction to directional statistics

Directional statistics is a branch of statistics dealing with observations that are directions. In most cases, these observations lie on the circumference of the unit circle of $\mathbb{R}^2$ (one then speaks of circular statistics) or on the surface of the unit hypersphere of $\mathbb{R}^p$ for $p \geq 3$ (implying the terminology of spherical statistics)[1]. Data of this type typically arise in meteorology (wind directions), astronomy (directions of cosmic rays or stars), earth sciences (location of an earthquake's epicentre on the surface of the earth) and biology (circadian rhythms, studies of animal navigation), to cite but these. The key difficulty when dealing with such data is the curvature of the sample space since the unit hypersphere or circle is a non-linear manifold. This can readily be seen on a very simple example. Imagine two points on the sphere of $\mathbb{R}^3$ and consider their average. This point will in general not lie on the sphere. This reasoning of course extends to several points on the sphere, entailing that the basic concept of sample mean needs to be adapted in order to yield a true mean direction on the sphere. Thus, the wheel has to be reinvented for virtually every classical concept from multivariate statistics.

This state of mind has, however, been ignored for a long time. Primitive statistical analysis of directional data can be traced back to the beginning of the 19th century by the likes of C. F. Gauss, yet it took till the seminal paper by Fisher (1953) before researchers became fully aware of the necessity to take the curved nature of the sample space into account. Sir Ronald Fisher successfully laid out the consequences of simplifying every directional problem through linear approximations. Quoting Fisher (1953): "*Any topological framework requires the developments of a theory of errors of characteristic and appropriate mathematical form.*" Indeed, while it was perfectly reasonable that Gauss used a tangent space approximation

[1]This includes data on the torus (product of two circles or spheres) and cylinder (product of $\mathbb{R}^p$ with a circle or sphere), but not other more general manifolds such as Stiefel or Grassmann manifolds.

for his highly concentrated (directional) astronomical measurements, the same cannot be said about more dispersed datasets. Fisher used as illustrating example the direction of remanent magnetism found in igneous or sedimentary rocks.

The impact of the Fisher (1953) paper led to a methodological and systematic study of directional data holding account of their actual topology. Numerous procedures and directional distributions were proposed and studied, mostly by extending to the directional setting classical concepts from multivariate analysis such as point estimation, one- and multi-sample testing procedures, or regression. For detailed accounts of these early developments from distinct perspectives, we refer the reader to the monographs by Mardia (1972), Watson (1983), Fisher et al. (1987) and Fisher (1993), to the discussion paper by Mardia (1975) as well as to the review paper by Jupp & Mardia (1989).

After this very active period, there was almost a lull in directional research in the 1990s. The situation has fortunately changed since the beginning of the new millennium. We have identified three main reasons for this resurgence of interest. First, the highly influential and very comprehensible book by Mardia & Jupp (2000) rendered directional statistics very popular. The ease of exposition combined with the versatility of presented results attracted the interest of theoretical statisticians as well as practitioners. Moreover, one year later appeared the seminal book by Jammalamadaka & SenGupta (2001) focussing on circular statistics. Second, and partially a consequence of the previous argument, researchers from life sciences, ecology and machine learning, among others, recognized the importance of directional statistics for their works. This has led to new demands and hence the necessity of novel methods and procedures. Third and finally, the technological advances have reshaped the entire field of statistics. The exponential increase in computing power and the availability of massive amounts of data have brought computer-intensive methods and high-dimensional statistics to the forefront of modern research in statistics, and hence also to directional statistics.

1.1.2 A brief outline of the theoretical advances presented in this book

The present book covers important theoretical developments in directional statistics over the past two decades, more precisely since the cornerstone reference books by Mardia & Jupp (2000) and Jammalamadaka & SenGupta (2001). We perceive it as a natural complement to these monographs, emphasizing modern research in the field without repeating the material covered in earlier books. The book is meant to be self-contained. We provide in Section 1.3 the basic knowledge required to follow the exposition of material throughout the monograph. The chapters all start with an introduction to the topic and briefly mention the state-of-the-art before the year 2000.

Numerous research themes in directional statistics have been addressed in recent years, resulting in a vast number of relevant methods and studies. We have decided to build our book on the following five pillars:

- Flexible parametric modeling: most classical distributions are able to model the center of a data cloud and the concentration around that center. Aspects such as skewness, peakedness or multi-modality cannot be addressed with these models. Flexible modeling means the search for versatile parametric distributions able to capture directional data features beyond location and concentration. This domain has flourished since 2005.

- Non-parametric inference: non-parametric statistics refer here mostly to kernel density estimation and rank-based inference. First defined at the end of the 1980s for spherical data, kernel density estimates have been largely studied, extended to other settings (torus and cylinder) and refined over the past ten years. Rank-based inferential procedures have a long-standing history on $\mathbb{R}^p$ but only appeared recently in directional statistics.

- Asymptotic statistics: most inferential procedures, also for directional data, rely on asymptotic results. We shall describe crucial concepts such as contiguity, local asymptotic normality and Le Cam's theory of asymptotic experiments. The latter is a complex but highly useful methodology to build efficient statistical methods based on solid mathematical statistics grounds. Initially designed for linear statistics, it was extended to the spherical setting in 2013 by the authors and their coauthors. We shall further discuss the nowadays unavoidable (n, p)-asymptotics where both the sample size n and the dimension p of the data are very large. This is intrinsically linked to the next point, namely

- High-dimensional directional statistics: high-dimensional statistics figure among the hottest topics in contemporary statistics. Numerous datasets have dimensions larger than the sample size, thwarting the validity of existing methods. This challenge has been taken up over the past decade, in particular driven by machine learning and genetics applications.

- Computational and graphical methods: the incredible increase in both computing power and complexity of the data have led to new handles for directional data. One the one hand, new visualization techniques reveal important underlying data structures. On the other hand, efficient and computationally fast algorithms allow dealing with complex datasets such as high-dimensional data or order-restricted data.

These ideas will be expounded in six thematic chapters whose detailed content is given in Section 1.4. Of course, there are several further topics we had envisaged to cover. We imposed on ourselves the above restriction for the sake of clarity and to avoid overlaps with other recent books and the companion book *Applied Directional Statistics: Modern Methods and Case Studies*, hereafter MMCS, for which we act as editors. The stochastic properties of needlets, a new form of spherical wavelets, as well as their applications to cosmic microwave background radiation data are reviewed in Chapter 10 of Marinucci & Peccati (2011). The even more recent monograph by Pewsey et al. (2013) is dedicated to a detailed account of how to implement circular statistics methods in R. The growing field of spatial directional statistics will be covered in MMCS. Modern application domains presented in MMCS will include bioinformatics, machine learning, cosmology, ecology, environmental sciences and behavioral sciences.

1.2 Directional datasets

We now want to further familiarize the reader with directional statistics by describing five types of directional datasets. These are taken from completely different domains and most of them reflect the aforementioned modern application areas. They have thus been a driving force behind several of the theoretical developments that we present in the subsequent chapters.

1.2.1 Paleomagnetism

We start with the classical domain of application that underpinned the argument of Fisher (1953) to take into account the correct topology of the data: paleomagnetism. This term refers to the study of the Earth's magnetic field in rocks, sediments, lava flows or other archeological materials. Certain minerals indeed conserve the direction of magnetic field when they cool down, allowing geologists to gain information about the Earth's magnetic field from ancient times as well as about plate tectonics. The associated data are spherical by nature, as illustrated in Figure 1.1. This figure represents measurements of remanent magnetization in red slits and claystones made at two different locations in Eastern New South Wales, Australia.

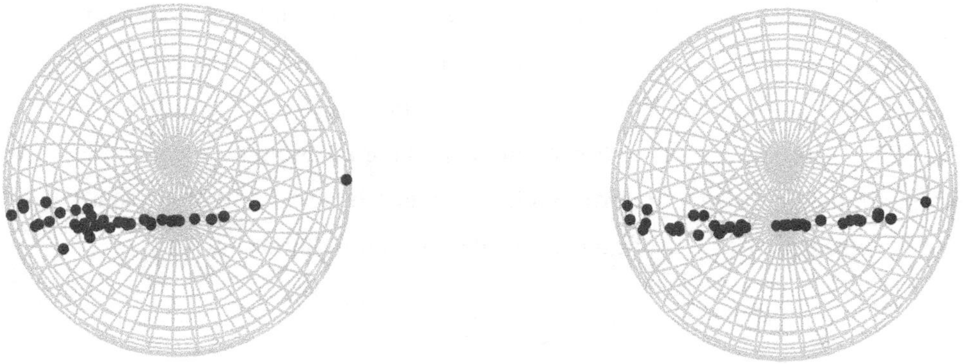

Figure 1.1: Measurements of remanent magnetization in red slits and claystones made at two different locations in Australia.

Typical research questions related to such data are the estimation of their modal direction, estimation of the concentration around that direction, related hypothesis tests about specific values, the determination of a probability distribution fitting the data or a graphical device to illustrate the main features of the data. We describe several modern methods dealing with these issues throughout this monograph. In the special situation of Figure 1.1, that is, when two or more datasets at different locations are involved, new information can be gathered about the above-mentioned tectonic plate movements or about whether the magnetization has been acquired before some deformation. In plain words, one is interested in the question: "do measurements of remanent magnetization at different locations come from the same source?" This problem is rather old in paleomagnetism. It was popularized in the seminal paper by Graham (1949) who developed the fold test for paleomagnetic data. In mathematical terms, this problem becomes a hypothesis testing problem. Suppose that we have m distinct datasets spread around sources $\boldsymbol{\mu}_i$, $i = 1, \ldots, m$, where each source $\boldsymbol{\mu}_i$ lies on the unit sphere of $\mathbb{R}^3$. Then the question becomes an ANOVA testing problem of the form $\mathcal{H}_0 : \boldsymbol{\mu}_1 = \boldsymbol{\mu}_2 = \cdots = \boldsymbol{\mu}_m$ against $\mathcal{H}_1 : \exists 1 \leq i \neq j \leq m$ such that $\boldsymbol{\mu}_i \neq \boldsymbol{\mu}_j$. Several papers have addressed this issue, and in recent years the underlying assumptions have been lowered thanks to new theoretical advances. We refer in particular to Chapter 5.

1.2.2 Political sciences

Circular data are not only measured via a compass, but also via a clock. Be it the 24-hour clock or the 12-months period, it is very convenient to represent time data on a circle if the end points (0.00 a.m. and 12.00 p.m., or January 1 and December 31) are naturally linked. This gain of information was recently recognized by political scientists (Gill & Hangartner 2010). In Figure 1.2 we show a rose diagram of gun crimes committed in Pittsburgh, Pennsylvania, between January 1, 1992 and May 31, 1996. The data are recorded in intervals of an hour and represent the time when an incident with a firearm (murder, robbery, assault, etc.) was reported. As can be expected, the peak of gun crimes lies around midnight.

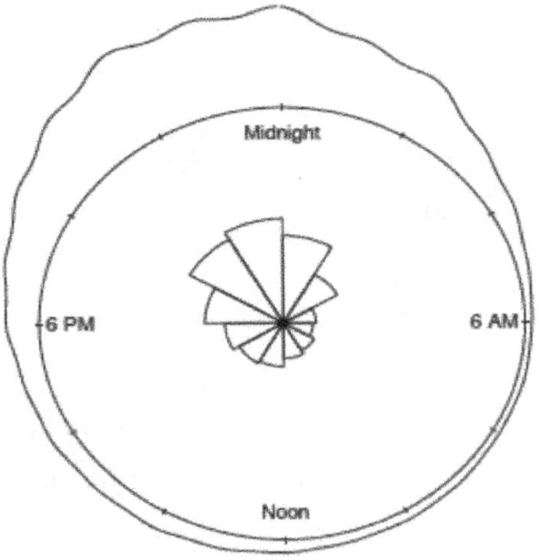

Figure 1.2: Rose diagram of gun crimes committed in Pittsburgh, Pennsylvania, and measured at the hourly level on the 24-hour clock.

The gun crimes dataset raises several questions. Is the data unimodal, and if so, symmetric about its mode? Which distribution describes best the gun crimes data? Answers can be obtained through the methods described in the subsequent chapters, especially Chapter 2 which contains various choices for parametric distributions that may fit the data from Figure 1.2.

1.2.3 Text mining

A popular branch of machine learning is text mining, where the goal is to categorize a variety of texts according to the similarity of their contents. This is achieved by considering an ensemble of words appearing neither too regularly nor too rarely in texts, and by counting the number of occurrences of each word in each text. One thus attributes to every observation, i.e., text, a vector with dimension the cardinality of the ensemble of words. In statistical terms, the categorization then corresponds to a cluster analysis of the resulting vectors. Empirical evidence suggests that one should normalize the vectors in order to remove potential categorization bias due to different text lengths. Think, for instance, of two texts where the second is obtained by copy-pasting and aligning ten times the first text. Their information content is exactly the same, but the word vectors will have highly distinct lengths. It is therefore the direction of the vector that is relevant for the cluster analysis.

Now, these directional word vectors are typically high-dimensional. Banerjee et al. (2005) considered *inter alia* Classic3, a collection of 3893 documents among which 1400 are from aeronautical system papers, 1033 from medical papers, and 1460 from information retrieval papers. The related word vector consists of 4666 words. Other text sources used by Banerjee et al. (2005) stem from Yahoo News and the CMU Newsgroup.

These text mining data thus are both high-dimensional and directional by nature, inducing the need for appropriate methods to analyze them. Such methods are displayed in Chapter 7, and the difficulty of parameter estimation in high dimensions is briefly addressed in Chapter 4. We shall not discuss here the relevant directional clustering algorithm as this problem will be addressed in the companion book *Applied Directional Statistics: Modern Methods and Case Studies.*

1.2.4 Wildfire orientation

Fire ecology is a branch of ecology that focuses on the origins of wildfires and tries to assess relationships with the surrounding environment. Mediterranean type climate countries are especially concerned with wildfires. The landscape changes in Portugal, for instance, are mainly driven by large and devastating wildfires. Barros et al. (2012) subdivided the map of Portugal into 102 watersheds and suggested the existence of preferential fire perimeter orientation at the level of these watersheds; see Figure 1.3. Their dataset spans over 31 years (1975–2005) and consists of the orientation as well as the size (in hectares) of each wildfire. The orientation can be considered at both the two-dimensional and three-dimensional level, yielding circular and spherical data. Combined with the fire size, which is a positive real value, this leads to cylindrical data.

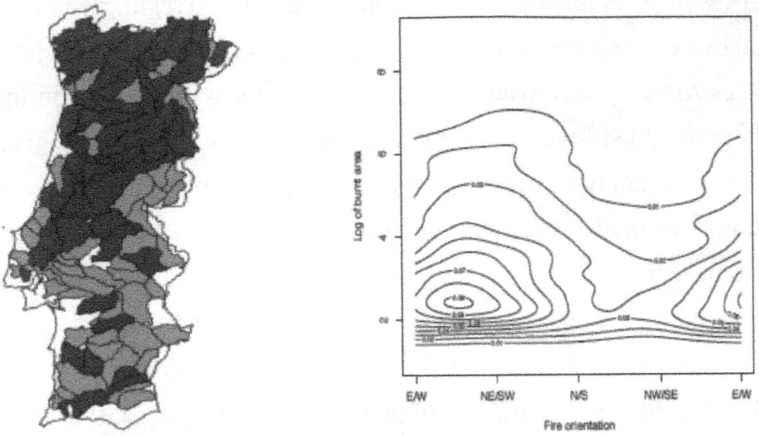

Figure 1.3: Left: map of wildfires in Portugal with the 102 watersheds identified by Barros et al. (2012). The dark regions are watersheds where the fires have preferential alignment. Right: contour plot for fires in a watershed with 1543 fires; the fires' orientation has a preferential alignment with the watershed orientation. We thank the authors from García-Portugués et al. (2014) for allowing us to use their pictures.

The main question of interest in the present context is to know whether there exists a relationship between fire orientation and fire size, as this would provide further insight into the study of wildfires. A non-parametric independence test is described in Chapter 3. Figure 1.3 provides a contour plot of orientation-size data for a given watershed. An interesting challenge is to find a parametric distribution incorporating this joint behavior.

1.2.5 Life sciences and bioinformatics

Predicting the correct three-dimensional structure of a protein on the basis of its one-dimensional protein sequence is a fundamental problem in life sciences. Solving this holy grail problem would have wide-reaching consequences in drug discovery, biotechnology and evolutionary biology, for instance. Nowadays massive databases of DNA and protein sequences are available, and structural bioinformatics is the domain within bioinformatics concerned with the prediction of the associated three-dimensional structure.

A protein consists of a sequence of amino acids, which essentially defines a protein's three-dimensional shape and dynamic behavior. Mathematically, the structure is for many purposes adequately described using dihedral angles (assuming ideal bond lengths and bond angles). The global shape of the protein, in particular the backbone structure, can be parameterized using the ϕ and ψ dihedral angles, between certain atoms of amino acids; see Figure 1.4. The vast existing database of known protein structures allows drawing scatter plots of the pair (ϕ, ψ) and hence knowing what configurations are more likely than others. This scatter plot bears the name Ramachandran plot (Ramachandran et al. 1963) and is illustrated in Figure 1.4. In recent years, it has been noted that it can be highly beneficial to develop probabilistic models of these angles in proteins. The data being two angles per amino acid, we are thus facing toroidal data structures.

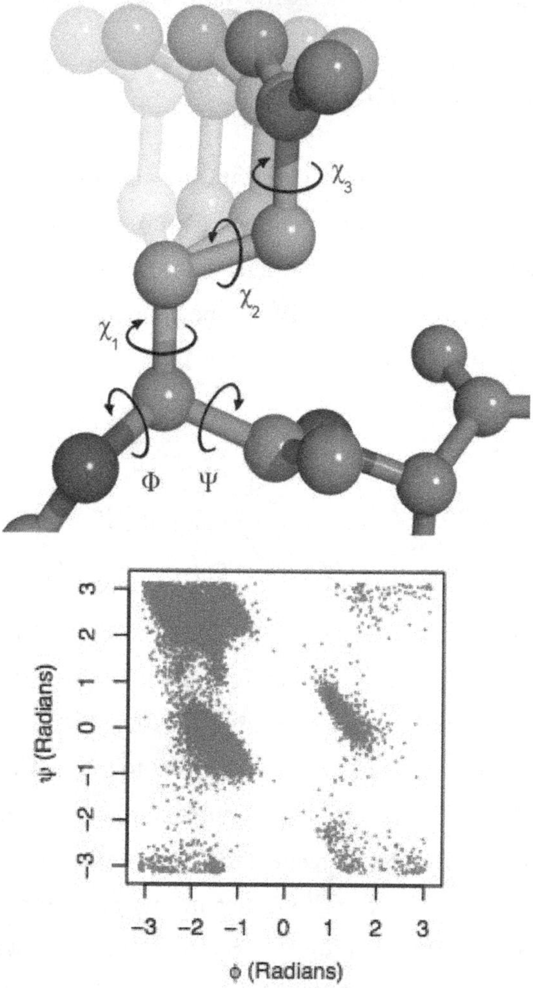

Figure 1.4: Top: Representation of the dihedral angles ϕ and ψ in glutamate. These angles are the main degrees of freedom for the backbone of an amino acid. The χ angles determine its side chain. Bottom: Example of a Ramachandran plot of the dihedral angles ϕ and ψ, expressed in radians. We thank Thomas Hamelryck for providing us with these pictures of which the former appeared in the paper Harder et al. (2010).

The most natural research question related to the protein data is to find a probability distribution matching the joint behavior of the ϕ and ψ angles, ideally conditional on the protein sequence. This can be achieved via some of the models presented in Chapter 2. Non-parametric approaches provide a distinct vision on the structure of proteins. Dealing with the dihedral angles is of course just one instance of the protein structure prediction problem. We shall not delve here into further details on this topic, which will be treated in the companion book *Applied Directional Statistics: Modern Methods and Case Studies*. For very detailed information, we refer the reader to the book by Hamelryck et al. (2012).

1.3 Basics and notations

In this section, we shall equip the reader with the necessary background to understand and appreciate the developments of the subsequent chapters. In particular, we shall explain recurrently used notations and abbreviations.

Points on the unit circle of $\mathbb{R}^2$ will not be represented as a two-dimensional vector $\mathbf{x} = (x_1, x_2)' \in \mathbb{R}^2$ with Euclidean norm $||\mathbf{x}|| = \sqrt{\mathbf{x}'\mathbf{x}} = 1$, but instead as an angle $\theta = \arctan^*(x_2/x_1)$ where

$$\arctan^*(x_2/x_1) := \begin{cases} \arctan(x_2/x_1) & \text{if} \quad x_1 \geq 0 \\ \arctan(x_2/x_1) + \pi & \text{if} \quad x_1 < 0, x_2 > 0 \\ \arctan(x_2/x_1) - \pi & \text{if} \quad x_1 < 0, x_2 \leq 0 \end{cases}$$

with arctan taking values in $[-\pi/2, \pi/2]$. In this definition we have arbitrarily fixed the origin of the circle at $(1, 0)'$ as well as its orientation as anti-clockwise; see Figure 1.5. A circular random variable thus can be perceived as a random angle over the interval $[-\pi, \pi)$ (or $[0, 2\pi)$), subject to the restriction that its density takes the same values at both endpoints. In other words, circular densities are simply defined as densities f (with respect to the Lebesgue measure) on the interval $[-\pi, \pi)$ for which $f(-\pi) = f(\pi)$. We opt in this book for the interval $[-\pi, \pi)$ instead of $[0, 2\pi)$ because then circular reflective symmetry can be nicely expressed as $f(-\theta) = f(\theta)\, \forall \theta \in [-\pi, \pi)$. Cumulative distribution functions F are defined as integrals between $-\pi$ and the point at which they are evaluated, and they further satisfy the important periodicity condition $F(z + 2\pi) - F(z) = 1$ for all $z \in \mathbb{R}$ (which means that any arc of length 2π has probability 1).

To provide the reader with a taste of the difficulties inherent to directional statistics, let us consider the very simple sample average of circular data points.

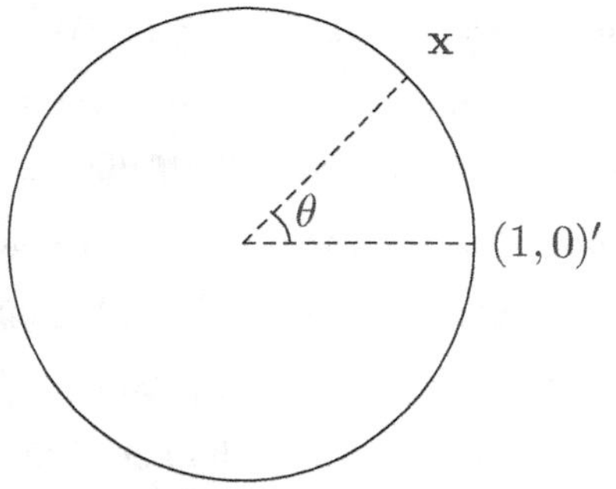

Figure 1.5: Representation of circular variables as angles measured anti-clockwise from the direction $(1, 0)'$.

The two most obvious definitions yield erroneous answers: the sample average of two points $\mathbf{x}_1 = (\cos(\theta_1), \sin(\theta_1))'$ and $\mathbf{x}_2 = (\cos(\theta_2), \sin(\theta_2))'$ can be obtained neither by considering the angular average $(\theta_1 + \theta_2)/2$ (e.g., the average of $-\pi + 0.01$ and $\pi - 0.01$ would yield 0 instead of $-\pi$) nor through the vector average $(\mathbf{x}_1 + \mathbf{x}_2)/2$ (the resulting point will nearly always lie inside the disc). Instead, the *mean direction* $\bar{\theta}$ of a set of angles $\theta_1, \ldots, \theta_n$ is defined as the angle related with the direction of $\sum_{i=1}^n \mathbf{x}_i$ and given by

$$\bar{\theta} = \arctan^* \left(\frac{\sum_{i=1}^n \sin(\theta_i)}{\sum_{i=1}^n \cos(\theta_i)} \right). \tag{1.1}$$

The concentration around the mean direction is measured by the *mean resultant length*

$$\bar{R} = n^{-1} \sqrt{\left(\sum_{i=1}^n \cos(\theta_i) \right)^2 + \left(\sum_{i=1}^n \sin(\theta_i) \right)^2}, \tag{1.2}$$

a quantity that we will often encounter on the following pages.

The (hyper-)sphere in $\mathbb{R}^p$ for $p \geq 3$ will be denoted $\mathcal{S}^{p-1} := \{\mathbf{z} \in \mathbb{R}^p : ||\mathbf{z}|| = 1\}$. Here we will not consider angles or spherical coordinates, but instead data points $\mathbf{x}$ on $\mathcal{S}^{p-1}$. This is why we shall throughout express densities on $\mathcal{S}^{p-1}, p \geq 3$, under the form $f(\mathbf{x})$ subject to $||\mathbf{x}|| = 1$. It is to be noted that we will express all our densities with respect to the usual surface area measure σ_{p-1} on $\mathcal{S}^{p-1}$, defined as

$$\sigma_{p-1}(\mathcal{S}^{p-1}) = \omega_p := \frac{2\pi^{p/2}}{\Gamma(p/2)}.$$

The *spherical mean* of points $\mathbf{x}_1, \ldots, \mathbf{x}_n$ is given by $\frac{\bar{\mathbf{x}}}{||\bar{\mathbf{x}}||}$ where $\bar{\mathbf{x}} = \frac{1}{n} \sum_{i=1}^n \mathbf{x}_i$. It is to be noted that the mean resultant length on hyper-spheres is conveniently given by $\bar{R} = ||\bar{\mathbf{x}}||$ (which is the spherical equivalent to (1.2)).

The two other directional supports with which we shall work, the torus and the cylinder, can easily be described with help of the above. The torus $\mathcal{T}^1$ is the product of two unit circles $\mathcal{S}^1 \times \mathcal{S}^1$ (or, for the hyper-torus $\mathcal{T}^{p-1}$, the product $\mathcal{S}^{p-1} \times \mathcal{S}^{p-1}$), while the cylinder $\mathcal{C}^1$ is the product of a unit circle and the real line $\mathcal{S}^1 \times \mathbb{R}$ (or, for the hyper-cylinder $\mathcal{C}^{p-1,k}$, the product $\mathcal{S}^{p-1} \times \mathbb{R}^k$ with integer k not necessary equal to $p-1$). In this book (Chapters 2 and 3), our goal with these supports shall mainly be to describe, parametrically and non-parametrically, their joint behavior, which is a non-trivial task.

The notations established above for random variables will not vary throughout the book. A random angle will be represented by the symbol Θ, a random vector on the sphere by $\mathbf{X}$, a random variable on the real line by Z and a random vector on $\mathbb{R}^p$ by $\mathbf{Z}$. The corresponding non-random quantities are denoted by $\theta, \mathbf{x}, z$ and $\mathbf{z}$, respectively. Moreover, location parameters shall invariably be represented by the symbol μ: μ for an angle, $\boldsymbol{\mu}$ for a point on the sphere, μ' for a point on the real line (except when no ambiguity is possible, in which case we shall also use simply μ on the real line).

We conclude this section by some general notations and remarks. The abbreviation iid stands for "independent and identically distributed". The $p \times p$ identity matrix is written $\mathbf{I_p}$ and a vector of, say p zeros, will be denoted $\mathbf{0} = (0, 0, \ldots, 0)' \in \mathbb{R}^p$. We shall often indicate the asymptotic behavior of functions or variables by having recourse to $o(\cdot)$ and $O(\cdot)$ quantities. Consider two functions $a(n)$ and $b(n)$. The notation $a(n) = o(b(n))$ as $n \to \infty$ means that $\frac{a(n)}{b(n)} \to 0$ as $n \to \infty$, while $a(n) = O(b(n))$ as $n \to \infty$ means that there exists $M > 0$ such that $\lim_{n\to\infty} \frac{a(n)}{b(n)} \leq M$. The notations $o_P(\cdot)$ and $O_P(\cdot)$ are used in the same way for convergence in probability of random variables. Finally, throughout the book we shall encounter special functions such as the modified Bessel function or the associated Legendre function. We shall briefly indicate a definition at their first occurrences, but refer the interested reader to the books by Abramowitz & Stegun (1965) and Gradshteyn & Ryzhik (2015) for more insights into special functions.

1.4 Plan of the book

The remainder of the book is organized as follows. Each of the six chapters starts by recalling the relevant notions from linear statistics on $\mathbb{R}^p$ before delving into the directional statistics themes, and ends with a "Further reading" section. All references are given in the bibliography at the very end of the book. Before describing the plan of the book in more detail, we would like to attract the reader's attention to an important point. When passing from Chapter 4 to Chapter 5, the tone will change as the presented material will become more mathematically involved. The second half of the book is strongly oriented towards hypothesis testing on directional supports.

Chapter 2 provides a very detailed description of the recently proposed probability distributions for data on the circle, sphere, torus and cylinder. The quest for flexible models that do fit the often complex aspects in modern datasets has culminated in a plethora of new distributions, especially on the circle. We hope to provide the reader with a guidance through this jungle of densities by trying to underline the respective merits and drawbacks of each proposal. To this end, we start with a brief recollection of the classical models like the von Mises, cardioid and wrapped Cauchy densities, and underline their limitations. This chapter on flexible models is a smooth start to the book and equips the reader with the necessary distributional knowledge to read the subsequent chapters.

Chapter 3 is in some sense the opposite of Chapter 2 as it deals with non-parametric density estimation. Kernel density estimators are well-established tools in linear statistics, and they were extended to the sphere in the late 1980s. Refinements of these proposals, especially with respect to the choice of the bandwidth parameter, have been suggested over the past years. Moreover, non-parametric density estimators were designed for toroidal and cylindrical data. We shall retrace these developments and amend them with related non-parametric inferential procedures such as goodness-of-fit tests, independence tests and regression.

Chapter 4 deals with computational and graphical methods. It consists of four distinct topics: ordering on the sphere, inference under order restrictions on the circle, non-parametric exploratory data analysis on the circle, and computationally fast parameter estimation for high-dimensional Fisher–von Mises–Langevin distributions. Each topic is reported in a separate section.

- Ordering data on the sphere is performed by means of spherical quantiles and depth functions. As will be shown, these intuitive notions allow the construction of QQ-plots, DD-plots and goodness-of-fit tests. The visualization of spherical quantiles provides new insight into the concentration of the data around their center.

- Ordering points around the circle forms the basis of order-restricted inference. Motivated by examples from genomics (phase angles of cell-cycle genes) and psychology, the aim is to estimate several angles under a given order restriction on these parameters. This is achieved by adapting the classical PAVA algorithm to the circular setting.

- Non-parametric exploratory data analysis on the circle is a graphical addendum to the non-parametric density estimation theory of Chapter 3. It describes a tool, the CircSiZer, that allows assessing visually the main features of a data set like the number of modes. As the name suggests, the CircSiZer is a circular extension of the popular SiZer on $\mathbb{R}$.

- The prevalence of high-dimensional datasets entails computational intricacies. In particular, estimating the concentration parameter of FvML distributions becomes very tricky. Its maximum likelihood estimate, despite being available in closed form, involves complex functions and hence requires approximations. The state-of-the-art approximations from Mardia & Jupp (2000) however cease to be valid in high-dimensional settings. We shall retrace the path of constantly improving approximations developed over the past years, and discuss their computational costs.

Chapter 5 exposes the adaptation of Le Cam's theory of asymptotic experiments from $\mathbb{R}^k$ to the spherical setting, a research stream initiated *inter alia* by the authors. In order to familiarize the reader with the topic, we begin with a detailed description of the classical linear Le Cam theory and illustrate the respective steps through a red-thread example. We then move to the spherical adaptation and show how this new theory permits us to build asymptotically optimal yet robust inferential procedures. In particular, we consider signed-rank-based estimation and hypothesis testing for the spherical location parameter, ANOVA on spheres and asymptotic power calculations.

Chapter 6 is a pure hypothesis testing chapter. More precisely, it considers recent advances in two classical topics: testing for uniformity and symmetry on the sphere (and circle). On the one hand, we will show how the well-known Rayleigh test of uniformity enjoys optimality features, discuss Sobolev and random projection type tests of uniformity, and treat the very delicate issue of testing uniformity in the presence of noisy data. On the other hand, we will present optimal semi-parametric tests for reflective symmetry on the circle and for rotational symmetry on the sphere, where optimality is reached against classes of skew distributions introduced in Chapter 2. We then slightly deviate from the main topic of the chapter by reconsidering the spherical location problem from a new angle, namely in the vicinity of uniformity. All procedures involving optimality features rely on the Le Cam methodology from Chapter 5.

Finally, Chapter 7 addresses high-dimensional situations. We start by introducing distributions on high-dimensional spheres, hereby complementing Chapter 2. Next we show how to extend classical tests to the high-dimensional setting. In particular, we consider uniformity, location and concentration tests. The different convergence regimes of the tests are particularly salient: certain tests require that the dimension p grows as a certain function of the sample size n, while other tests are free from such restrictions. The latter are said to be (n, p)-universally valid, a highly desirable situation in practice. We conclude the chapter with the description of principal nested spheres, an elegant dimension-reduction technique mimicking principal component analysis.

Advances in flexible parametric distribution theory

2.1 Introduction

2.1.1 Flexible parametric modeling: an active research area on $\mathbb{R}^p$

Probability distributions are the building blocks of several branches of statistics, among which are data modeling, hypothesis testing, regression and time series analysis. The normal, exponential, Student t, beta and log-normal are popular examples of distributions on the real line or on subintervals of $\mathbb{R}$. In higher dimensions, the multivariate normal and the class of elliptically symmetric distributions on $\mathbb{R}^p$, $p \geq 2$, have been thoroughly studied and are nowadays classical textbook examples (Fang et al. 1990).

Since more and more data tend to exhibit non-trivial characteristics such as skewness, varying tailweight or multimodality, there is an increasing need for distributions able to capture these features. Such distributions are usually referred to as *flexible*, and we shall adopt that terminology here. Popular instances of flexible models on $\mathbb{R}^p$, developed since the end of the nineteenth century, include

- the skew-symmetric distributions popularized by Adelchi Azzalini in the 1980s, resulting from multiplying a symmetric density by a so-called skewing function (Azzalini & Capitanio 2014);

- the two-piece distributions, mainly used in the scalar case, resulting from separating a symmetric distribution at its center, introducing distinct scale parameters on either side and glueing together the two halves (Wallis 2014);

- the transformation-of-variables distributions, resulting from the transformation of a random variable/vector by means of a monotone increasing diffeomorphism (Ley & Paindaveine 2010a);

- mixture distributions, resulting from a convex combination of two or more unimodal distributions (McLachlan & Peel 2000).

These distributions have been successfully used to model datasets from domains such as finance, economics, environmetrics or biometrics. Flexible distributions on $\mathbb{R}^p$ are nowadays so numerous that categorizations, such as that above, have become essential. We refer to the recent review papers by Jones (2015) and Ley (2015) for details, many further distributions and a plethora of references on this active research area.

2.1.2 Organization of the remainder of the chapter

The quest for flexible distributions has also become an important research topic in directional statistics over the past two decades. This chapter is devoted to modern advances in flexible modeling on directional supports such as the circle (Section 2.2), the torus and cylinder (Section 2.4) and the hypersphere (Section 2.3). As the reader will notice, the major part of this chapter is dedicated to circular distributions, as they are by far the most studied.

2.2 Flexible circular distributions

2.2.1 Four ways to construct circular densities

The simplest example of a distribution on the circle is the *uniform distribution*, with density $f_U(\theta) = \frac{1}{2\pi}$ over $[-\pi, \pi)$. A reader unfamiliar with directional statistics may find it *a priori* difficult to define non-uniform distributions on the circle, which is why we start by describing four general ways to construct circular densities.

- *Wrapping approach*: a distribution on $\mathbb{R}$ is wrapped around the circumference of the unit circle. Letting Z be a random variable with density f_Z on $\mathbb{R}$, the wrapped-f circular random variable $\Theta \in [-\pi, \pi)$ is defined via $\Theta = (Z + \pi) \pmod{2\pi} - \pi$ and its density is given by

$$f_\Theta(\theta) = \sum_{k=-\infty}^{\infty} f_Z(\theta + 2\pi k). \tag{2.1}$$

  For instance, if f_Z is the normal density, one obtains the well-known *wrapped normal distribution*. The latter, like the majority of known wrapped distributions on the circle, suffers however from a major drawback: (2.1) does not simplify to a closed form. As a consequence, wrapped circular densities are often not easy to handle, the best known exception being the wrapped Cauchy distribution; see Section 2.2.2.

- *Conditioning approach*: express a distribution on $\mathbb{R}^2$ as the joint distribution of the polar coordinates length and angle (r, Θ), and consider then the distribution of the angle conditionally on the restriction $r = 1$. A well-known example is the von Mises distribution (see Section 2.2.2), obtained by conditioning a bivariate normal distribution with mean $(\cos(\mu), \sin(\mu))'$ and covariance matrix

$$\begin{pmatrix} \kappa^{-1} & 0 \\ 0 & \kappa^{-1} \end{pmatrix},$$

  $\mu \in [-\pi, \pi), \kappa > 0$.

- *Projection approach*: a distribution on $\mathbb{R}^2$ is radially projected onto the unit circle. In terms of polar coordinates, the length part is integrated out to obtain the marginal density of the angular part. A popular case of this approach is the projected normal distribution; see Section 3.5.6 of Mardia & Jupp (2000).

A common feature of the three approaches is that they build upon distributions on $\mathbb{R}$ or $\mathbb{R}^2$. In contrast to this, the fourth approach is more direct:

- *Perturbation approach*: a circular density is multiplied by a function under the constraint that the resulting product is again a proper circular density. The cardioid distribution (see Section 2.2.2) is a classical example of this approach.

2.2.2 The classics: von Mises, cardioid and wrapped Cauchy distributions

We shall now briefly present three popular circular distributions. All three distributions are obtained by one of the constructions described in the previous section, all three are characterized by two parameters, a location parameter $\mu \in [-\pi, \pi)$ and a non-negative parameter regulating the concentration of the distribution around μ, all three contain the circular uniform distribution as special case and, except for that special case, all three are symmetric about their unique mode μ. This notion of symmetry on the circle is called *reflective symmetry*, but we shall in this chapter simply speak of symmetry.

The *von Mises distribution* has density

$$\theta \mapsto \frac{1}{2\pi I_0(\kappa)} \exp(\kappa \cos(\theta - \mu)), \tag{2.2}$$

with I_0 the modified Bessel function of the first kind and of order 0,[1] and concentration $\kappa \geq 0$. It plays a core role among circular distributions. It is hence not surprising that Gumbel et al. (1953) refer to it as the Circular Normal distribution. Initially derived by von Mises (1918) via a maximum likelihood characterization (see Section 2.3.2), numerous distinct generations of the von Mises density have been proposed in the literature, among which is the conditioning approach mentioned in the previous section. We refer the reader to Section 3.5.4 of Mardia & Jupp (2000) and Section 2.2.4 of Jammalamadaka & SenGupta (2001) for a thorough study of the density (2.2).

The *cardioid distribution* has density

$$\theta \mapsto \frac{1}{2\pi}\left(1 + 2\rho\cos(\theta - \mu)\right), \tag{2.3}$$

with concentration $0 \leq \rho < 1/2$. We readily see that the cardioid density is a perturbation of the uniform density by multiplication with $(1 + 2\rho\cos(\theta - \mu))$. Jeffreys (1948), page 302, introduced this circular distribution, whose properties are described in Section 3.5.5 of Mardia & Jupp (2000).

Finally, the *wrapped Cauchy distribution* has density

$$\theta \mapsto \frac{1}{2\pi}\frac{1 - \ell^2}{1 + \ell^2 - 2\ell\cos(\theta - \mu)}, \tag{2.4}$$

with concentration $0 \leq \ell < 1$. It stands out as one of the rare wrapped densities with closed form. Its origins can be traced back to Lévy (1939). Mardia & Jupp (2000) discuss various properties of the wrapped Cauchy distribution in their Section 3.5.7.

The von Mises, cardioid and wrapped Cauchy distributions contain the uniform as a special case, obtained for $\kappa = \rho = \ell = 0$, respectively. The effects of the concentration parameters and the differences between the three distributions can be appreciated from a consideration of Figure 2.1. We direct the reader's attention to the fact that properties such as trigonometric moments or the mean resultant length for the three models can be deduced from the general results of Section 2.2.4. We conclude this succinct introduction to the circular classics by referring the reader

[1]The modified Bessel function of the first kind and of order $\alpha \geq 0$, $I_\alpha(z)$ for $z > 0$, admits the integral representation

$$I_\alpha(z) = \frac{1}{\pi}\int_0^\pi \exp(z\cos(\theta))\cos(\alpha\theta)d\theta - \frac{\sin(\alpha\pi)}{\pi}\int_0^\infty \exp(-z\cosh(y) - \alpha y)dy.$$

This formula readily shows why the modified Bessel function of the first kind and order 0, $I_0(z)$, appears in the normalizing constant of the von Mises density on the circle.

once again to the monographs by Mardia & Jupp (2000) and Jammalamadaka & SenGupta (2001) for inferential aspects related to these distributions.

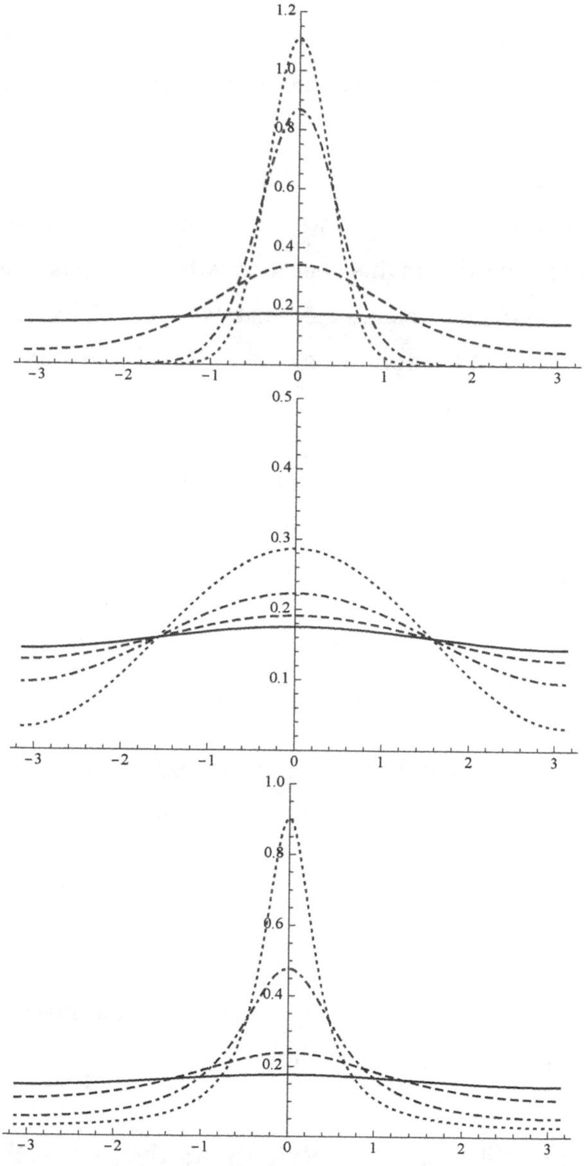

Figure 2.1: Top: von Mises densities with $\kappa = 0.1, 1, 5, 8$. Middle: cardioid densities with $\rho = 0.05, 0.1, 0.2, 0.4$. Bottom: wrapped Cauchy densities with $\ell = 0.05, 0.2, 0.5, 0.7$. In each figure, the location is zero and increasing values of the concentration correspond to the solid, dashed, dash-dotted and dotted lines.

2.2.3 Beyond the classics: modern flexible circular modeling

The von Mises, cardioid and wrapped Cauchy distributions are symmetric and unimodal and their shapes are determined by a single concentration parameter. They share these properties with other known distributions such as the wrapped normal. Consequently, such densities cannot capture data features such as asymmetry, bimodality or multimodality.

In response to these needs, various flexible models for circular data have been proposed over the past decade, some of which we shall highlight in the following sections. An important question in this context is: what properties should a "good" flexible circular model possess, besides the capacity to model diverse distributional shapes? We have identified the following list of desirable features for a flexible family of circular distributions:

- The density should be of a tractable form and amenable to calculations. This is crucial when one wishes to determine, for instance, the probability of having data occurring far away from the mode.
- The number of parameters should be as small as possible and the parameters should have clear interpretations.
- Parameter estimation should be straightforward: usually maximum likelihood or method of moments estimation methods are used.
- The family should nest well-known sub-models, as this permits model reduction.

In the subsequent sections, we shall analyze various recent proposals of flexible circular models on the basis of those requirements.

2.2.4 Flexible modeling of symmetric data: the Jones–Pewsey distribution

The seminal paper by Jones & Pewsey (2005) introduced a general three-parameter family of symmetric circular distributions that contains several well-known distributions from the literature. Its density is given by

$$\theta \mapsto \frac{(\cosh(\kappa\psi))^{1/\psi}}{2\pi P_{1/\psi}(\cosh(\kappa\psi))}\,(1+\tanh(\kappa\psi)\cos(\theta-\mu))^{1/\psi}, \qquad (2.5)$$

where $P_\alpha(\cdot)$ is the associated Legendre function of the first kind of degree $\alpha \in \mathbb{R}$ and order 0,[2] $\mu \in [-\pi, \pi)$ is the location parameter, $\kappa \geq 0$ the concentration parameter and $\psi \in \mathbb{R}$ is a shape parameter termed the shape index by Jones and Pewsey. When $\kappa = 0$ or $\psi \to \pm\infty$ (with κ finite), (2.5) is the uniform density. The choices $\psi = 1$ and -1 respectively yield the cardioid and wrapped Cauchy density, while the limit $\psi \to 0$ yields the von Mises density. When $\psi > 0$ and $\kappa \to \infty$, the Jones–Pewsey density (2.5) corresponds to Cartwright's power-of-cosine density (Cartwright 1963), whereas for $-1 < \psi < 0$ it becomes the circular t-density of Shimizu & Iida (2002). We omit shapes of (2.5) here, as they can be seen as a continuous variation between the densities shown in Figure 2.1. The normalizing constant of the circular Jones–Pewsey family of Section 2.2.4 is given by

$$\int_{-\pi}^{\pi} (1 + \tanh(\kappa\psi)\cos(\theta))^{1/\psi} d\theta \quad =$$

$$\frac{1}{\cosh^{1/\psi}(\kappa\psi)} \int_{-\pi}^{\pi} (\cosh(\kappa\psi) + \sinh(\kappa\psi)\cos(\theta))^{1/\psi} d\theta = \frac{2\pi P_{1/\psi}(\cosh(\kappa\psi))}{\cosh^{1/\psi}(\kappa\psi)},$$

where the last equality holds for both positive and negative ψ.

The wide range of the symmetric Jones–Pewsey family is not the only reason for its popularity. Further distributional trumps are its unique mode μ for all $\psi \in \mathbb{R}$ and all $\kappa > 0$, a normalizing constant expressed in terms of known functions, and stochastic representations as conditional distributions of spherically/elliptically symmetric distributions on $\mathbb{R}^2$ (see Section 2.5 of Jones & Pewsey 2005).

To simplify the presentation of the characteristic function and trigonometric moments, we fix $\mu = 0$. The sine moments $\beta_k = \mathrm{E}[\sin(k\Theta)]$ are all zero due to the symmetry of the Jones–Pewsey density. Consequently, the characteristic function $\{\phi_k = \mathrm{E}[e^{ik\Theta}] : k = 0, \pm 1, \ldots\}$, with Θ following the Jones–Pewsey density (2.5), is defined via the cosine moments $\alpha_k = \mathrm{E}[\cos(k\Theta)]$ given by

$$\alpha_k = \begin{cases} \frac{\Gamma(1/\psi+1) P^k_{1/\psi}(\cosh(\kappa\psi))}{\Gamma(1/\psi+k+1) P_{1/\psi}(\cosh(\kappa\psi))} & \text{if } \psi > 0, \\ \frac{I_k(\kappa)}{I_0(\kappa)} & \text{if } \psi = 0, \\ \frac{\Gamma(1/|\psi|-k) P^k_{1/\psi}(\cosh(\kappa\psi))}{\Gamma(1/|\psi|) P_{1/\psi}(\cosh(\kappa\psi))} & \text{if } \psi < 0, \end{cases}$$

where $\Gamma(\cdot)$ denotes the gamma function and $P^\beta_\alpha(\cdot)$ the associated Legendre func-

[2]For $z \in [1, \infty)$, the integral representation of the associated Legendre function $P_\alpha(z)$ corresponds to

$$P_\alpha(z) = \frac{1}{2\pi} \int_{-\pi}^{\pi} \left(z + \sqrt{z^2 - 1}\cos(\theta)\right)^\alpha d\theta.$$

tion of the first kind of degree $\alpha \in \mathbb{R}$ and order $\beta \in \mathbb{R}^+$.[3] Of particular interest is the (population) mean resultant length

$$\rho = \alpha_1 = \frac{|\psi|}{(1+\psi)} \frac{P^1_{1/\psi}(\cosh(\kappa\psi))}{P_{1/\psi}(\cosh(\kappa\psi))}.$$

It increases with concentration κ, while there is no monotone effect in ψ.

Concerning parameter estimation, no closed-form expressions exist for the maximum likelihood (ML) estimators of the parameters of the Jones–Pewsey density, and numerical optimization methods are required. Provided that $|\kappa\psi| < 6$, the profile log-likelihood surface for these two parameters (obtained by maximizing the likelihood with respect to μ) is concave with a clearly defined maximum, leading to rapid convergence of the numerical maximization procedure. Instabilities only arise for $|\kappa\psi|$ too big. It is notable that the ML estimate for the location parameter is asymptotically independent of the ML estimates of the other two parameters, which are correlated. For more details, see Section 3 of Jones & Pewsey (2005), as well as the appendix of that paper where the elements of the observed and expected information matrix are provided.

2.2.5 Sine-skewing: a simple tool to skew any symmetric distribution

In order to model asymmetric data on the circle, Umbach & Jammalamadaka (2009) and Abe & Pewsey (2011*a*) proposed a simple means to skew any symmetric distribution, thereby exploiting the richness of existing symmetric circular laws. Their construction consists in perturbing a density f symmetric about $\mu \in [-\pi, \pi)$ into

$$\theta \mapsto f(\theta - \mu)(1 + \lambda \sin(\theta - \mu)), \tag{2.6}$$

where $\lambda \in (-1, 1)$ plays the role of a skewness parameter. When $\lambda = 0$, no perturbation occurs and the base symmetric density is retained, otherwise (2.6)

[3] The associated Legendre function of the first kind, $P^\beta_\alpha(z)$ for $z \in [1, \infty)$, with degree $\alpha \in \mathbb{R}$ and order $\beta \in \mathbb{R}^+$ admits the integral representations

$$\begin{aligned} P^\beta_\alpha(z) &= \frac{(-\alpha)_\beta}{\pi} \int_0^\pi \frac{\cos(\beta\theta)}{(z + \sqrt{z^2-1}\cos(\theta))^{\alpha+1}} d\theta \\ &= \frac{\Gamma(\alpha+1)}{\pi\Gamma(\alpha-\beta+1)} \int_{-\pi}^0 \frac{\cos(\beta\theta)}{(z - \sqrt{z^2-1}\cos(\theta))^{\alpha+1}} d\theta, \end{aligned}$$

where

$$(-\alpha)_\beta = (-1)^\beta \frac{\Gamma(\alpha+1)}{\Gamma(\alpha-\beta+1)}.$$

is skewed to the left ($\lambda > 0$) or to the right ($\lambda < 0$). Abe and Pewsey term such densities *sine-skewed* because they are obtained by perturbing f with $(1+\lambda\sin(\theta-\mu))$ (recall the perturbation approach discussed in Section 2.2.1). A sine-skewed-f random variable Θ_{SSF} with location $\mu = 0$ is generated via the following very simple mechanism:

1. Generate a circular random variable Θ_F with density f symmetric about 0.

2. Generate independently a uniform random variable U on $[0, 1]$.

3. Define Θ_{SSF} as follows:

$$\Theta_{SSF} = \begin{cases} \Theta_F & \text{if} \quad U \leq (1+\lambda\sin(\Theta_F))/2 \\ -\Theta_F & \text{if} \quad U > (1+\lambda\sin(\Theta_F))/2. \end{cases}$$

A clear advantage of sine-skewing is that the normalizing constant is unaffected. Trigonometric moments are expressed in terms of the moments of f; see Abe & Pewsey (2011*a*) for details. A drawback lies in the potential bimodality of sine-skewed densities. The strictly unimodal sine-skewed wrapped Cauchy distribution is a notable exception. As an illustration of this fact, we provide in Figure 2.2 the plots of sine-skewed von Mises, cardioid and wrapped Cauchy densities. It is also worth noting that, for f the uniform density, (2.6) becomes the cardioid density with mode at $(\mu + \pi/2 + \pi) \pmod{2\pi} - \pi$.

Concerning inferential properties, Abe & Pewsey (2011*a*) warn that method of moments estimators will not always exist. Maximum likelihood estimation requires numerical maximization, for which Abe and Pewsey recommend using a global optimization routine based on a differential evolution algorithm instead of the Nelder–Mead simplex method, as the latter may yield local maxima. Further inferential properties of sine-skewed densities can be found in Section 6.6 of Chapter 6, which describes tests for symmetry ($\lambda = 0$) that are optimal against sine-skewed alternatives.

Umbach & Jammalamadaka (2009) also suggested a more general skewing method that transforms the symmetric density f into

$$\theta \mapsto 2f(\theta - \mu)G(\omega(\theta - \mu)) \tag{2.7}$$

where $G(\theta) = \int_{-\pi}^{\theta} g(y)dy$ is the cumulative distribution function (cdf) of a given circular symmetric density g and ω is a weighting function satisfying for all $\theta \in [-\pi, \pi)$ the three conditions $\omega(-\theta) = -\omega(\theta)$, $\omega(\theta + 2\pi k) = \omega(\theta)\, \forall k \in \mathbb{Z}$, and

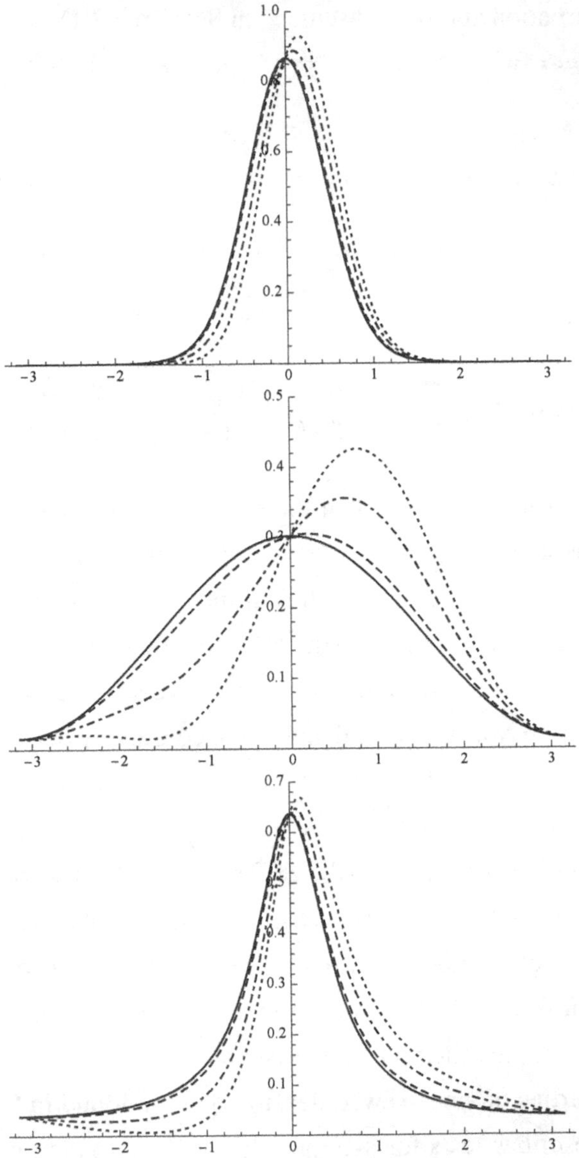

Figure 2.2: Sine-skewed versions of the von Mises density with $\kappa = 5$ (top), cardioid density with $\rho = 0.45$ (middle) and wrapped Cauchy density with $\ell = 0.6$ (bottom), for $\lambda = 0, 0.1, 0.5$ and 0.9. In each figure, increasing values of λ correspond to the solid, dashed, dash-dotted and dotted lines.

$|\omega(\theta)| \leq \pi$. Choosing $G(\theta) = (\pi + \theta)/(2\pi)$, the cdf of the circular uniform distribution, and $\omega(\theta) = \lambda\pi\sin(\theta)$ with $\lambda \in (-1, 1)$ obviously leads to (2.6).

The densities (2.6) and (2.7) are the circular analogs of the skew-symmetric distributions on $\mathbb{R}^p$ mentioned in Section 2.1.1.

2.2.6 Skewness combined with unimodality: the scale-transforming approach

In the previous section we learnt that sine-skewing can produce bimodal skew distributions. The goal of Jones & Pewsey (2012) was to avoid this by proposing skew densities that remain unimodal. Their construction turns any symmetric circular density f with mode $\mu = 0$ (again, for simplicity of presentation only) into its *scale-transformed* version via

$$f(\theta) \stackrel{\tau}{\Rightarrow} f(\tau(\theta)) \tag{2.8}$$

for some monotone scaling function τ. The idea underpinning (2.8) is not new and can be traced back to the late 1970s with Papakonstantinou (1979) and Batschelet (1981). The two transformations considered in those papers are

$$\tau_{\nu_1}(\theta) = \theta + \nu_1 \sin(\theta) \quad \text{and} \quad \tau_{\nu_2}(\theta) = \theta + \nu_2 \cos(\theta)$$

with $\nu_1, \nu_2 \in \mathbb{R}$. While τ_{ν_1} maintains symmetry but affects peakedness (rendering f either more flat-topped or more sharply peaked), τ_{ν_2} skews any symmetric density. Both τ_{ν_1} and τ_{ν_2} preserve unimodality when $-1 \leq \nu_1, \nu_2 \leq 1$. Papakonstantinou applied the two transformations to the cardioid density, Batschelet to the von Mises density. In recent years, these transformations have seen a resurgence of interest. Abe et al. (2009) provided an in-depth study of Papakonstantinou's τ_{ν_1}-scaled cardioid density, while Pewsey et al. (2011) did the same for Batschelet's τ_{ν_1}-scaled von Mises density. Abe et al. (2013) studied the effects of both transformations in considerable generality.

Let us now come back to the Jones & Pewsey (2012) paper and their goal of a simple means of obtaining a skew unimodal density. Their idea consists in slightly extending τ_{ν_2} into

$$\tau_{2,\nu}(\theta) = \theta - \nu - \nu\cos(\theta), \quad -1 \leq \nu \leq 1,$$

and then considering $\tau_{\nu,JP}(\theta) = \tau_{2,\nu}^{-1}(\theta)$. This allows turning any symmetric density f into

$$f(\tau_{\nu,JP}(\theta)) = f\left(\tau_{2,\nu}^{-1}(\theta)\right) \tag{2.9}$$

with skewness parameter ν and explains why Jones and Pewsey speak of *inverse Batschelet distributions*. Densities (2.9) are unimodal; we provide plots of $\tau_{\nu,JP}$-transformed von Mises, cardioid and wrapped Cauchy densities in Figure 2.3. In contrast to the original proposals τ_{ν_1} and τ_{ν_2}, $\tau_{\nu,JP}$ leaves the normalizing constant unchanged, sharing this advantage with sine-skewed densities. There exists, in fact, a direct link between both skewing methods: if Θ follows (2.9), then $\tau_{\nu,JP}(\Theta)$ generates a random variable following the corresponding sine-skewed f density (2.6). Since sine-skewed densities are simple to generate (see Section 2.2.5), this link implies that inverse Batschelet densities inherit this property. Deepening the comparison between inverse Batschelet transforming and sine-skewing, the former leads in general to higher degrees of asymmetry, while the latter is more amenable to calculations such as trigonometric moments.

Parameter estimation for scale-transformed or inverse Batschelet densities is performed via maximum likelihood, Jones and Pewsey suggesting the Nelder–Mead algorithm. Of course, the inverse operation in (2.9) slows down the maximisation procedure. These densities enjoy a remarkable inferential property when applied to a three-parameter symmetric distribution (such as the Jones–Pewsey, for instance), namely parameter orthogonality between a block formed by the location and skewness parameter and another block formed by the concentration and additional shape parameter. This property entails that, asymptotically, the ML estimate of concentration, for example, has no influence on the behavior of the skewness ML estimate. This rare property mimicks two-piece distributions on the real line, see Jones & Anaya-Izquierdo (2011).

2.2.7 A general device for building symmetric bipolar distributions

Bimodal data have often been modeled by means of two-component mixtures of unimodal distributions. Symmetric bipolar data are bimodal data symmetric about the two modes μ_1 and μ_2 which are diametrically opposed, that is, separated by π radians. Such data arise in a variety of natural sciences such as meteorology and animal behavior experiments.

Abe & Pewsey (2011*b*) proposed a general way to model symmetric bipolar data. Starting from a unimodal symmetric density $f(\theta - \mu)$ with mode at $\mu \in [-\pi, \pi)$, their two-step procedure proceeds as follows:

1. Duplicate $f(\theta - \mu)$ into $f(2(\theta - \mu))$, an antipodally symmetric density (meaning that the density takes the same values for opposite points on the

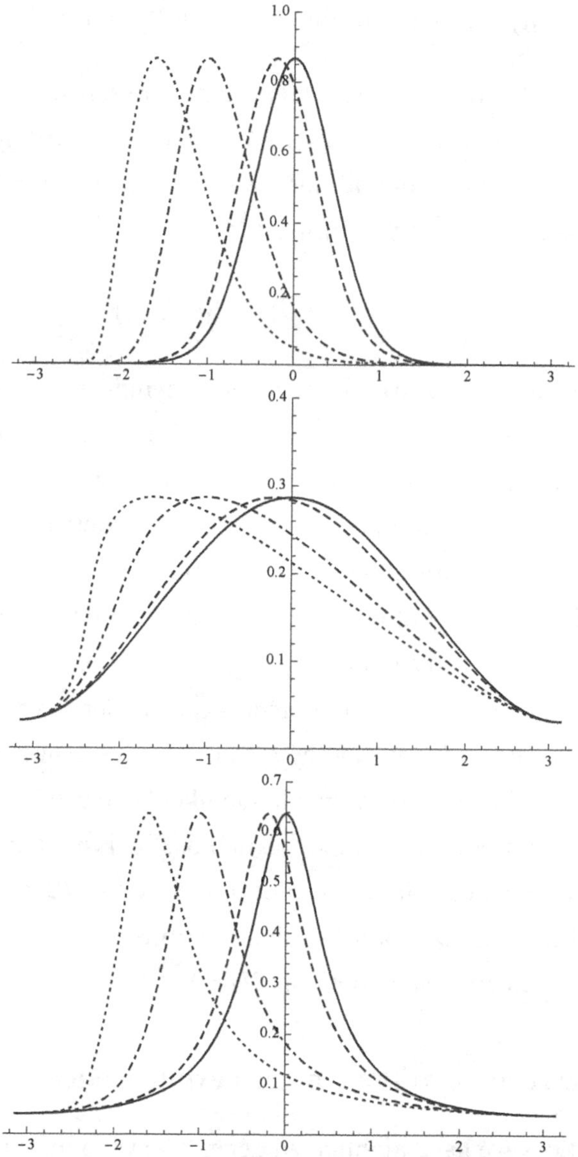

Figure 2.3: Scale-transformed (inverse Batschelet) versions of the von Mises density with $\kappa = 5$ (top), cardioid density with $\rho = 0.4$ (middle) and wrapped Cauchy density with $\ell = 0.6$ (bottom), for $\nu = 0, 0.1, 0.5$ and 0.8. In each figure, increasing values of ν correspond to the solid, dashed, dash-dotted and dotted lines.

circle and sphere).

2. Perturb $f(2(\theta-\mu))$ by multiplication with $(1+\lambda\cos(\theta-\mu))$ for $\lambda \in [0,1]$.

The first step creates a bimodal density whose modes are separated by π radians, and the second step assigns different weights to both modes depending on the value of λ, which hence endorses the role of mode-weighting parameter. The resulting symmetric bipolar densities are of the form

$$\theta \mapsto (1+\lambda\cos(\theta-\mu))f(2(\theta-\mu)). \tag{2.10}$$

For $\lambda = 0$, densities of form (2.10) are antipodally symmetric, while increasing values of λ increase the probability mass around μ. This effect of λ is illustrated in Figure 2.4. The above construction, based on duplication and cosine perturbation, avoids the calculation of complex normalizing constants and yields simple expressions for the trigonometric moments as functions of the trigonometric moments of the base density f. A potential drawback lies in the fact that densities of form (2.10) can also be uni- and trimodal.

Parameter estimation via method of moments can be, depending on the initial density f, straightforward. Abe & Pewsey (2011*b*) suggest using the method of moments estimates as initial values for maximum likelihood estimation. However, an identifiability issue arises at $\lambda = 0$: the "main mode" μ is no longer identifiable (see (2.10)). In such a case one should restrict μ to lie in $[-\pi/2, \pi/2)$. Antipodal symmetry can readily be tested via a likelihood ratio test of the null hypothesis $\mathcal{H}_0 : \lambda = 0$ against the alternative hypothesis $\mathcal{H}_1 : \lambda \neq 0$.

2.2.8 A brief description of three other flexible models

In the previous sections we have highlighted general ways to introduce skewness, varying tail-weight and multimodality into symmetric unimodal distributions. In contrast, we shall now present a selection of flexible constructions that were tailor-made to extend a single targeted density.

The Generalized von Mises distribution

At first sight, the *Generalized von Mises (GvM) distribution* may seem out of place in our listing of "new" models, as it has a long history; relevant references include Maksimov (1967), Rukhin (1972) and Yfantis & Borgman (1982). However, it still

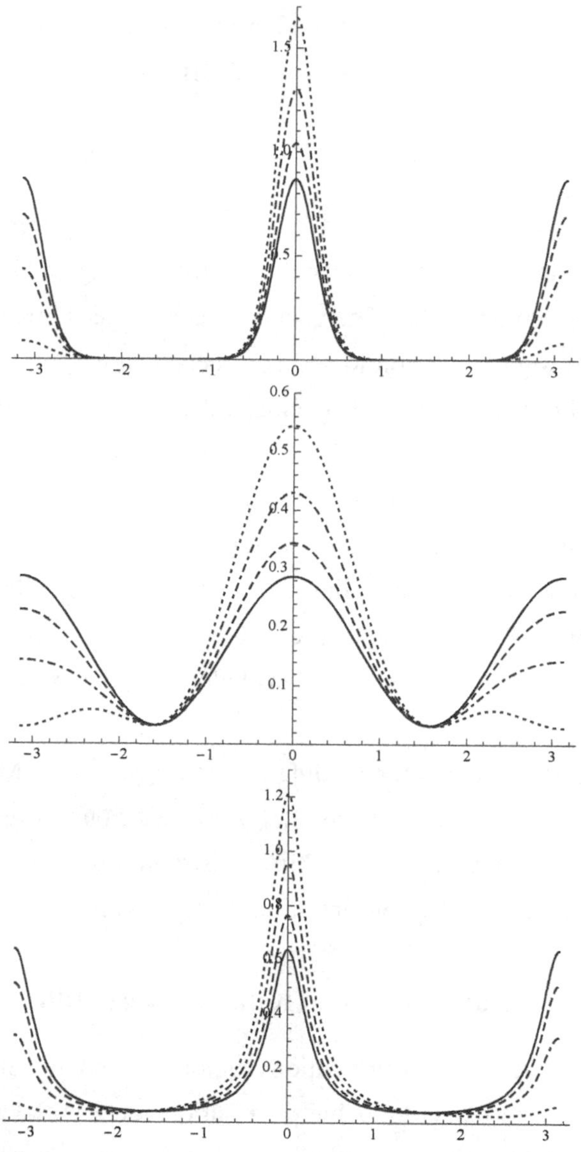

Figure 2.4: Duplicated and cosine perturbated versions of the von Mises density with $\kappa = 5$ (top), cardioid density with $\rho = 0.4$ (middle) and wrapped Cauchy density with $\ell = 0.6$ (bottom), for $\lambda = 0, 0.2, 0.5$ and 0.9. In each figure, increasing values of λ correspond to the solid, dashed, dash-dotted and dotted lines.

occupies an important role in modern distribution theory, as reflected by the references Gatto & Jammalamadaka (2007), Gatto (2008) and Gatto (2009) who have further studied the GvM. The GvM distribution is built from a bivariate normal distribution on $\mathbb{R}^2$ by a conditioning argument and has a density proportional to

$$\theta \mapsto \exp\left(\kappa_1 \cos(\theta - \mu_1) + \kappa_2 \cos(2(\theta - \mu_2))\right), \tag{2.11}$$

where $\mu_1, \mu_2 \in [-\pi, \pi)$ are location and $\kappa_1, \kappa_2 \geq 0$ concentration parameters. Such a construction can be considered as multiplicative mixing and is particularly suitable for modeling bimodal data. The GvM can capture asymmetry as well as symmetry, and obviously has the von Mises as a special case. The conditions for unimodality/bimodality depend on the behavior of the roots of a quartic equation. The normalizing constant is rather convoluted, and must be computed numerically or estimated via truncated series expansions as provided in Gatto (2009). Trigonometric moments, consequently, suffer from the same drawback.

Maximum likelihood estimation works very well for GvM distributions, as the log-likelihood function is concave and hence leads to a unique maximum. This nice feature follows from the fact that the GvM belongs to the exponential family of distributions.

We attract the reader's attention regarding terminology. The GvM distribution with density (2.11) is also referred to as the GvM_2 distribution (e.g., in Gatto 2009), a special case of Generalized von Mises distributions of order k, GvM_k, with k locations $\mu_1, \ldots, \mu_k$ and k concentrations $\kappa_1, \ldots, \kappa_k$.

The Möbius-transformed distributions of Kato & Jones (2010)

Applying monotone increasing transformations to normal random variables is one of the oldest approaches to create flexible distributions on the real line. On the circle, one transformation has raised particular interest in recent years: the Möbius transformation. It transforms an angle $\bar{\Theta}$ into Θ via the mapping

$$\bar{\Theta} \mapsto \Theta := \mu + \nu + 2\arctan\left(\omega_r \tan\left(\frac{1}{2}(\bar{\Theta} - \nu)\right)\right) \tag{2.12}$$

or

$$e^{i\Theta} = e^{i\mu} \frac{e^{i\bar{\Theta}} + re^{i\nu}}{re^{i(\bar{\Theta}-\nu)} + 1},$$

where $-\pi \leq \mu < \pi, 0 \leq \nu < 2\pi, 0 \leq r < 1$ are the three parameters of the transformation, with $\omega_r = (1-r)/(1+r)$ and complex number $i = \sqrt{-1}$. Kato

& Jones (2010) introduced a new four-parameter family on the circle by applying the Möbius transformation (2.12) to a random variable following a von Mises distribution centered at 0 and with concentration $\kappa \geq 0$. The resulting density is given by

$$\theta \mapsto \frac{1-r^2}{2\pi I_0(\kappa)} \exp\left[\frac{\kappa\left(\xi\cos(\theta-\eta)-2r\cos(\nu)\right)}{1+r^2-2r\cos(\theta-\gamma)}\right]\frac{1}{1+r^2-2r\cos(\theta-\gamma)}, \tag{2.13}$$

where $\gamma = \mu + \nu$, $\xi = \sqrt{r^4 + 2r^2\cos(2\nu) + 1}$ and

$$\eta = \mu + \arg\left(r^2\cos(2\nu) + 1 + ir^2\sin(2\nu)\right).$$

It contains as special cases the von Mises ($r = 0$), the wrapped Cauchy ($\kappa = 0$) as well as the uniform ($r = \kappa = 0$). Advantages of this Kato–Jones proposal are the simple normalizing constant and closure under Möbius transformation; note that the wrapped Cauchy is itself a Möbius-transformed uniform distribution. Parameter interpretation, however, is less clear-cut. While μ plays the role of a location parameter, it is the interplay between ν, r and κ that controls the skewness of the density. For instance, (2.13) is symmetric if and only if $\nu = 0, \nu = \pi, r = 0$ or $\kappa = 0$. Concentration and peakedness are mostly determined by κ and r. As for the Generalized von Mises distribution, density (2.13) can be both uni- and bimodal, with the conditions for unimodality depending on the discriminant of a quartic equation involving the three parameters ν, r and κ.

Regarding inferential issues, it seems from Kato & Jones (2010) that uniqueness of maxima of the log-likelihood is not guaranteed and they hence suggest multiple restarts of the optimization algorithm. An important bonus of the Möbius-transformed density (2.13) is that it lends itself well to circular-circular regression problems.

We conclude by remarking that the Möbius transformation can, more generally, be applied to other symmetric base densities, such as for instance the Jones–Pewsey density as briefly mentioned in Section 8 of Kato & Jones (2010), but this would result in a less parsimonious five-parameter model. Wang & Shimizu (2012) studied the Möbius-transformed cardioid distribution.

The very flexible unimodal distribution of Kato & Jones (2015)

Adopting a completely different approach, Kato & Jones (2015) proposed yet another four-parameter family based on the wrapped Cauchy distribution. Recalling the notation from (2.4), the trigonometric moments of the wrapped Cauchy density are given by $\phi_k^{\mathrm{WC}} = (\ell e^{i\mu})^k$ for non-negative k and $\phi_k^{\mathrm{WC}} = \overline{\phi_{-k}^{\mathrm{WC}}}$ for negative k, with $\bar{z}$ being the complex conjugate of z. Kato and Jones extend this expression to

$$\phi_k^{\mathrm{KJ15}} = \gamma(\ell e^{i\eta})^{-1}\left(\ell e^{i(\mu+\eta)}\right)^k, \quad k = 1, 2, \ldots, \tag{2.14}$$

with $\gamma \geq 0$ and $-\pi \leq \eta < \pi$. The wrapped Cauchy moments are retrieved when $\gamma = \ell$ and $\eta = 0$. Starting from this expression, there exists an absolutely continuous density on the circle whose trigonometric moments equal (2.14) if and only if $\mu, \eta \in [-\pi, \pi)$ and $\ell, \gamma \in [0, 1)$ satisfy

$$(\ell\cos(\eta) - \gamma)^2 + (\ell\sin(\eta))^2 \leq (1 - \gamma)^2, \tag{2.15}$$

and the corresponding density is of the form

$$\theta \mapsto \frac{1}{2\pi}\left(1 + 2\gamma\frac{\cos(\theta - \mu) - \ell\cos(\eta)}{1 + \ell^2 - 2\ell\cos(\theta - \mu - \eta)}\right). \tag{2.16}$$

In addition to the wrapped Cauchy, the cardioid density is also a special case of (2.16), obtained when $\ell = 0$. For $\gamma > 0$, the density is always unimodal, while $\gamma = 0$ yields the uniform distribution. This property is quite remarkable, as density (2.16) can be symmetric or asymmetric, flat-topped or sharply peaked. Moreover, the parameters bear clear interpretations: μ is the mean direction and γ is the mean resultant length. The circular skewness and kurtosis of Batschelet (1981), defined respectively as the imaginary and real parts of $\phi_2^{\mathrm{KJ15}}\exp(-2i\mu)$, are conveniently given by $cS = \gamma\ell\sin(\eta)$ and $cK = \gamma\ell\cos(\eta)$. It is hence possible to reparametrize the density in terms of the parameters cS and cK so as to have pleasing parameter interpretability in terms of location, concentration, skewness and kurtosis. Random variable generation can be achieved by acceptance/rejection algorithms based on random variables generated from a wrapped Cauchy distribution.

Parameter estimation is possible via both method of moments and maximum likelihood. The method of moments is particularly appealing thanks to the simple trigonometric moments, and the resulting estimates are good starting values for the numerical maximization of the log-likelihood function. In order to avoid condition (2.15) on the parameters during the optimization procedure, a reparameterization is suggested in Section 5.2 of Kato & Jones (2015).

Finally, we mention that the same pair of authors derived a further tractable four-parameter density in Kato & Jones (2013), again based on the wrapped

Cauchy density but this time derived via Brownian motion. Their distribution can be symmetric or asymmetric, unimodal or bimodal, peaked or flat-topped. No inferential issues are considered in the paper.

2.3 Flexible spherical distributions

In contrast to the circle $\mathcal{S}^1$, new flexible models on the (hyper-)sphere are relatively scarce. In addition to the intrinsic complexity of $\mathcal{S}^{p-1}$ when $p > 2$, this dearth might be due to the fact that various spherical distributions were studied in the 1970s (e.g., mixture models). Consequently, only one recent flexible proposal will be described in detail (Section 2.3.3), and we place more emphasis on new developments and known results for rotationally symmetric distributions (Section 2.3.2), an important class of spherical densities that we will recurrently encounter throughout this book.

2.3.1 Classical spherical distributions

The best known and simplest spherical distribution is the uniform distribution with density $\frac{1}{\omega_p}$ over $\mathcal{S}^{p-1}$, where we recall that $\omega_p = 2\pi^{p/2}/\Gamma(p/2)$ is the surface area of $\mathcal{S}^{p-1}$. No other distribution is invariant under both rotation and reflection. All four constructions of Section 2.2.1 remain valid on hyper-spheres; see Section 9.1 of Mardia & Jupp (2000). Letting $\mathbf{x} \in \mathcal{S}^{p-1}$ be a point on the unit hypersphere, popular non-uniform directional densities are the

- *Fisher–von Mises–Langevin (FvML) density*

$$\mathbf{x} \mapsto \frac{\left(\frac{\kappa}{2}\right)^{p/2-1}}{2\pi^{p/2} I_{p/2-1}(\kappa)} \exp(\kappa \mathbf{x}'\boldsymbol{\mu}), \tag{2.17}$$

  with $I_{p/2-1}$ the modified Bessel function of the first kind and of order $p/2 - 1$ (see Section 2.2.2). Density (2.17) is unimodal about the location $\boldsymbol{\mu} \in \mathcal{S}^{p-1}$ and its concentration around the mode is regulated by $\kappa \geq 0$, the limit $\kappa = 0$ yielding the uniform distribution. For $p = 2$ we retrieve the von Mises (1918) density, for $p = 3$ the Fisher (1953) density and, for general p, the Langevin (1905*b*,*a*) densities; hence the terminology. Like the von Mises on the circle, the FvML distribution plays a central role in spherical statistics. A further similarity is its genesis by conditioning on a p-variate normal distribution.

- *Fisher–Bingham density*

$$\mathbf{x} \mapsto \frac{1}{a(\kappa, \mathbf{A})} \exp(\kappa \mathbf{x}'\boldsymbol{\mu} + \mathbf{x}'\mathbf{A}\mathbf{x}), \tag{2.18}$$

with location parameter $\boldsymbol{\mu} \in \mathcal{S}^{p-1}$, concentration $\kappa \geq 0$ and $\mathbf{A}$ a symmetric $p \times p$ matrix, and where $a(\kappa, \mathbf{A})$ is a normalizing constant. Without loss of generality, one can assume that $\mathrm{tr}(\mathbf{A}) = 0$ since $\mathbf{x}'\mathbf{x} = 1$. When $\mathbf{A}$ is the zero matrix, (2.18) reduces to (2.17), when $\kappa = 0$ the *Bingham density* (Bingham 1964) is obtained and for $p = 2$ it takes the form of a Generalized von Mises density (Section 2.2.8). Approximating the normalizing constant $a(\kappa, \mathbf{A})$ is a challenging task that several researchers have addressed, see, e.g., Kume & Wood (2005).

- *Kent density*, which corresponds to (2.18) under the additional condition that $\mathbf{A}'\boldsymbol{\mu} = \mathbf{0}$. This modification, proposed in Kent (1982), leads to oval density contours around $\boldsymbol{\mu}$, the matrix $\mathbf{A}$ being a shape parameter.

- *(Dimroth–Scheidegger–) Watson density*

$$\mathbf{x} \mapsto \frac{\Gamma(p/2)}{2\pi^{p/2} M\left(\frac{1}{2}, \frac{p}{2}, \kappa\right)} \exp(\kappa(\mathbf{x}'\boldsymbol{\mu})^2), \tag{2.19}$$

with location $\boldsymbol{\mu} \in \mathcal{S}^{p-1}$ and concentration $\kappa \in \mathbb{R}$, and where $M(1/2, p/2, \kappa)$ denotes the Kummer function.[4] For $\kappa > 0$, the Watson density is bipolar with maxima at $\pm\boldsymbol{\mu}$, while for $\kappa < 0$ it is concentrated around the great circle like girdle densities. Density (2.19) is invariant under reflections and hence is suited for modeling *axial data* (data that are axes, hence such that each direction is considered as equivalent to the opposite direction). This distribution was introduced independently by Dimroth (1962, 1963) and Watson (1965).

For properties of these distributions, especially those of the FvML, and further classical directional distributions, see Sections 9.3 and 9.4 of Mardia & Jupp (2000).

[4]An integral representation of the Kummer function $M(a, b, z)$ with $a, b, z \in \mathbb{R}$ is

$$M(a, b, z) = \frac{\Gamma(b)}{\Gamma(b-a)\Gamma(a)} \int_{-1}^{1} e^{t^2 z} t^{2a-1}(1-t^2)^{b-a-1} dt.$$

2.3.2 Rotationally symmetric distributions

Most of the classical densities mentioned in the previous section share the common feature that they are *rotationally symmetric* about their location $\boldsymbol{\mu} \in \mathcal{S}^{p-1}$. The distribution of a random $\mathbf{X} \in \mathcal{S}^{p-1}$ is said to be rotationally symmetric about $\boldsymbol{\mu}$ if $\mathbf{OX}$ is equal in distribution to $\mathbf{X}$ for any orthogonal $p \times p$ matrix $\mathbf{O}$ satisfying $\mathbf{O}\boldsymbol{\mu} = \boldsymbol{\mu}$. This gives rise to densities of the form

$$\mathbf{x} \mapsto c_{f_a,p} f_a(\mathbf{x}'\boldsymbol{\mu}), \tag{2.20}$$

where the *angular function* $f_a : [-1, 1] \to \mathbb{R}^+$ is absolutely continuous and $c_{f_a,p}$ is a normalizing constant. The terminology "angular function" reflects the fact that the distribution of $\mathbf{X}$ only depends on the angle (colatitude angle when $p = 3$) between $\mathbf{X}$ and $\boldsymbol{\mu}$. The FvML and Watson distributions are popular instances of this class of distributions, with their respective angular functions being $t \mapsto \exp(\kappa t)$ and $t \mapsto \exp(\kappa t^2)$, $t \in [-1, 1]$. Other popular rotationally symmetric distributions are the Arnold distribution with angular function $t \mapsto \exp(-\kappa|t|)$ (Arnold 1941), the Selby distribution with angular function $t \mapsto \exp(\pm\kappa(1 - t^2)^{1/2})$ (Selby 1964) and the Purkayastha distribution with angular function $t \mapsto \exp(-\kappa \arccos(t))$ (Purkayastha 1991), with $\kappa \geq 0$ each time playing the role of concentration parameter.

Over the past decade this list of distributions has been further extended. Shimizu & Iida (2002) defined Pearson Type VII and t-distributions on the sphere, Siew & Shimizu (2008) introduced generalized symmetric Laplace distributions, while García-Portugués (2013) proposed the directional Cauchy and skew-normal densities. Jones & Pewsey (2005) extended the Jones–Pewsey family (Section 2.2.4) to the directional case by replacing the cosine function in their density (2.5) with the scalar product $\mathbf{x}'\boldsymbol{\mu}$, and adapting the normalizing constant to the p-dimensional setup. Ley et al. (2013) introduced the directional linear, logarithmic, logistic and square-root distributions with respective angular functions $t \mapsto t + a$, $t \mapsto \log(t + a)$, $t \mapsto \frac{a \exp(-b \arccos(t))}{(1+a \exp(-b \arccos(t)))^2}$ and $t \mapsto \sqrt{t + a}$, with a and b constants chosen in such a way that each mapping qualifies as an angular function. The directional linear distribution is an extension of the circular cardioid distribution.

Numerous statistical procedures presuppose rotational symmetry of the data at hand. In astronomy, for example, this assumption is natural when the earth's rotation only allows one to know the colatitude of the emission direction of certain radiations, and not the exact direction. Such practical constraints are, however, rather rare, and rotationally symmetric distributions are most often assumed because they generalize the FvML distribution and nest several other popular directional distributions; this parallels elliptically symmetric distributions on $\mathbb{R}^p$ (Paindaveine 2012) which are a convenient extension of the multivariate normal distribution. A further, and by no means minor, appeal of rotationally symmetric distributions are their nice stochastic properties.

An essential structural decomposition of random vectors $\mathbf{X} \in \mathcal{S}^{p-1}$ is the *tangent-normal decomposition* in the direction $\boldsymbol{\mu} \in \mathcal{S}^{p-1}$:

$$\mathbf{X} = (\mathbf{X}'\boldsymbol{\mu})\boldsymbol{\mu} + (1 - (\mathbf{X}'\boldsymbol{\mu})^2)^{1/2}\mathbf{S}_{\boldsymbol{\mu}}(\mathbf{X}) \tag{2.21}$$

where the sign vector $\mathbf{S}_{\boldsymbol{\mu}}(\mathbf{X}) := (\mathbf{X} - (\mathbf{X}'\boldsymbol{\mu})\boldsymbol{\mu})/\|\mathbf{X} - (\mathbf{X}'\boldsymbol{\mu})\boldsymbol{\mu}\|$ is defined on the tangent space

$$\mathcal{S}^{p-1}(\boldsymbol{\mu}^{\perp}) := \{\mathbf{v} \in \mathbb{R}^p \mid \|\mathbf{v}\| = 1, \mathbf{v}'\boldsymbol{\mu} = 0\}.$$

Figure 2.5 illustrates this decomposition. In the case of rotationally symmetric distributions, it leads to a very nice and useful property, which will play an important role also in later chapters of this book.

Lemma 2.3.1 (Watson 1983) *Let the distribution of* $\mathbf{X}$ *be rotationally symmetric on* $\mathcal{S}^{p-1}$ *about* $\boldsymbol{\mu}$. *Then (i)* $\mathbf{X}'\boldsymbol{\mu}$ *and* $\mathbf{S}_{\boldsymbol{\mu}}(\mathbf{X}) = (\mathbf{X} - (\mathbf{X}'\boldsymbol{\mu})\boldsymbol{\mu})/\|\mathbf{X} - (\mathbf{X}'\boldsymbol{\mu})\boldsymbol{\mu}\|$ *are stochastically independent and (ii) the multivariate sign vector* $\mathbf{S}_{\boldsymbol{\mu}}(\mathbf{X})$ *is uniformly distributed on* $\mathcal{S}^{p-1}(\boldsymbol{\mu}^{\perp}) := \{\mathbf{v} \in \mathbb{R}^p \mid \|\mathbf{v}\| = 1, \mathbf{v}'\boldsymbol{\mu} = 0\}$.

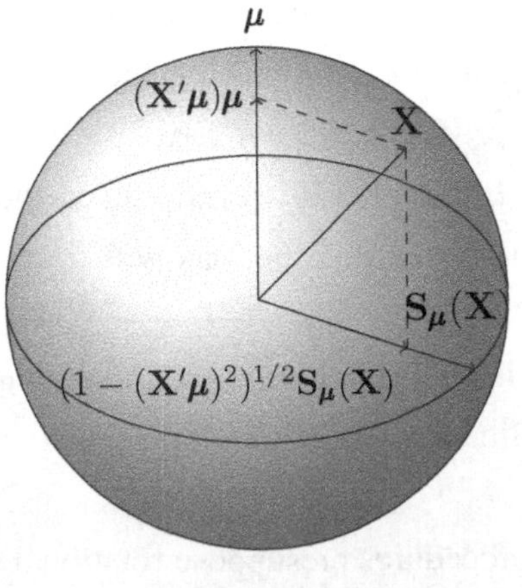

Figure 2.5: Illustration of the tangent-normal decomposition of a vector $\mathbf{X} \in \mathcal{S}^{p-1}$.

This decomposition of the sphere leads to the change of variables $d\sigma_{p-1}(\mathbf{x}) = (1-t^2)^{(p-3)/2} dt d\sigma_{p-2}(\mathbf{v})$ with $t \in [-1,1]$ and $\mathbf{v} \in \mathcal{S}^{p-2}$. With this in hand, one can readily see that the density of the projection $\mathbf{X}'\boldsymbol{\mu}$, for $\mathbf{X}$ rotationally symmetric around $\boldsymbol{\mu}$, is given by

$$t \mapsto \tilde{f}_a(t) := \frac{\omega_p \, c_{f_a,p}}{B(\frac{1}{2}, \frac{1}{2}(p-1))} f_a(t)(1-t^2)^{(p-3)/2}, \quad -1 \leq t \leq 1, \tag{2.22}$$

where $B(\cdot,\cdot)$ is the beta function. Indeed,

$$\begin{aligned}\int_{\mathcal{S}^{p-1}} c_{f_a,p} f_a(\mathbf{x}'\boldsymbol{\mu}) d\sigma_{p-1}(\mathbf{x}) &= \int_{\mathcal{S}^{p-2}} \int_{-1}^{1} c_{f_a,p} f_a(t)(1-t^2)^{(p-3)/2} dt d\sigma_{p-2}(\mathbf{v}) \\ &= \int_{-1}^{1} \omega_{p-1} c_{f_a,p} f_a(t)(1-t^2)^{(p-3)/2} dt \\ &= \int_{-1}^{1} \frac{\omega_p \, c_{f_a,p}}{B(\frac{1}{2}, \frac{1}{2}(p-1))} f_a(t)(1-t^2)^{(p-3)/2} dt,\end{aligned}$$

where the last equality follows from $\omega_p/\omega_{p-1} = B\left(\frac{1}{2}, \frac{p-1}{2}\right)$. A further appealing aspect of rotationally symmetric models is their belonging to statistical group models (Chang 2004).

We conclude this section with a characterization of rotationally symmetric distributions via their maximum likelihood estimators (MLEs), established in Duerinckx & Ley (2012). This property mimics MLE characterizations on $\mathbb{R}^p$. Indeed, a classical characterization result of Gauss (1809) states that the MLE of the parameter in a location family of distributions on the real line equals the sample mean for all samples $X_1, \ldots, X_n$ of all sample sizes n if and only if the samples are drawn from a Gaussian population. A similar characterization exists for the FvML distribution with respect to the spherical mean $\bar{\mathbf{X}}/||\bar{\mathbf{X}}||$: it was proved in the circular case by von Mises (1918), in the three-dimensional spherical setting by Arnold (1941) and Breitenberger (1963) and, for any dimension, in Bingham & Mardia (1975). This property is one of the major reasons why the FvML is considered to be the directional analogue of the normal distribution on $\mathbb{R}^p$. Duerinckx & Ley (2012) considered rotationally symmetric distributions in general. Defining, for $t \in [-1,1]$, the functions $H_{f_a}(t) := t\varphi_{f_a}\left(\sqrt{1-t^2}\right)$ and $\bar{H}_{f_a}(t) := t\varphi_{f_a}\left(-\sqrt{1-t^2}\right)$ with $\varphi_{f_a}(t) = f_a'(t)/f_a(t)$, they established the following result.

Theorem 1 (Duerinckx & Ley 2012) *Let f_a and g_a be two continuously differentiable angular functions on $[-1,1]$ associated with rotationally symmetric distri-*

butions on $\mathcal{S}^{p-1}$, $p \geq 2$, and write their respective location MLEs as $\hat{\boldsymbol{\mu}}_{f_a}$ and $\hat{\boldsymbol{\mu}}_{g_a}$. Suppose that H_{f_a} is invertible on $[-1, 1]$.

(i) Fix $N = 3 + \lceil (H_{f_a}(1))^{-1} \max_{t \in [0,1]} \tilde{H}_{f_a}(t) - 1 \rceil$. Then $\hat{\boldsymbol{\mu}}_{f_a} = \hat{\boldsymbol{\mu}}_{g_a}$ for all samples of fixed sample size $n \geq N$ if and only if there exist constants $c, d \in \mathbb{R}_0^+$ such that $g_a(t) = c(f_a(t))^d \, \forall t \in [-1, 1]$.

(ii) Fix $N = 3$ and suppose that f_a is such that φ_{f_a} is even. Then $\hat{\boldsymbol{\mu}}_{f_a} = \hat{\boldsymbol{\mu}}_{g_a}$ for all samples of fixed sample size $n \geq N$ if and only if there exist constants $c, d \in \mathbb{R}_0^+$ such that $g_a(t) = c(f_a(t))^d \, \forall t \in [-1, 1]$.

This theorem states which rotationally symmetric distributions can be characterized through their location MLE and provides the minimal conditions (in terms of required sample size and regularity of angular functions) under which this characterization holds.

2.3.3 A general method to skew rotationally symmetric distributions

In order to model asymmetrically distributed data on the sphere, Ley & Verdebout (2017) propose a general means of skewing any rotationally symmetric distribution. Let $c_{f_a,p} f_a(\mathbf{x}'\boldsymbol{\mu})$ be a rotationally symmetric density about $\boldsymbol{\mu} \in \mathcal{S}^{p-1}$ and $\Pi : \mathbb{R} \to [0, 1]$ a skewing function, that is, a monotone increasing continuous function satisfying $\Pi(-y) + \Pi(y) = 1$ for all $y \in \mathbb{R}$ (for instance, Π can be the cdf of any symmetric density on $\mathbb{R}$). We write $\boldsymbol{\Upsilon}_{\boldsymbol{\mu}}$ for the semi-orthogonal matrix such that

$$\boldsymbol{\Upsilon}_{\boldsymbol{\mu}} \boldsymbol{\Upsilon}'_{\boldsymbol{\mu}} = \mathbf{I}_p - \boldsymbol{\mu}\boldsymbol{\mu}' \quad \text{and} \quad \boldsymbol{\Upsilon}'_{\boldsymbol{\mu}} \boldsymbol{\Upsilon}_{\boldsymbol{\mu}} = \mathbf{I}_{p-1},$$

where $\mathbf{I}_p$ is the $p \times p$ identity matrix. The Π-skewed version of $c_{f_a,p} f_a(\mathbf{x}'\boldsymbol{\mu})$ according to Ley & Verdebout (2017) is then defined as

$$\mathbf{x} \mapsto 2c_{f_a,p} f_a(\mathbf{x}'\boldsymbol{\mu}) \Pi(\boldsymbol{\lambda}' \boldsymbol{\Upsilon}'_{\boldsymbol{\mu}} \mathbf{x}) \tag{2.23}$$

with $\boldsymbol{\lambda} \in \mathbb{R}^{p-1}$ a skewness parameter. The base symmetric density corresponds to $\boldsymbol{\lambda} = \mathbf{0}$, whilst non-zero values of $\boldsymbol{\lambda}$ produce skewed spherical densities. The motivation behind density (2.23) is to break rotational symmetry at the level of the uniformly distributed (over $\mathcal{S}^{p-1}(\boldsymbol{\mu}^{\perp})$) sign vector $\mathbf{S}_{\boldsymbol{\mu}}(\mathbf{X})$, see Lemma 2.3.1. Ley & Verdebout (2017) refer to densities of the form (2.23) as *skew-rotationally-symmetric densities*, by analogy with the skew-symmetric distributions on $\mathbb{R}^p$ (Azzalini & Capitanio 2014). They are a natural spherical extension of the sine-skewed circular densities presented in Section 2.2.5. As in the circular setting, the normalizing constant is unaffected, which is an even stronger asset in the case of hyper-spheres. Skew-rotationally-symmetric data can be readily generated thanks to a simple stochastic representation, see Ley & Verdebout (2017) for details. As an illustration, we have generated $n = 100$ observations from skewed versions of the FvML distribution; see Figure 2.6. Inferential procedures related to skew-rotationally-symmetric distributions are discussed in Section 6.7.

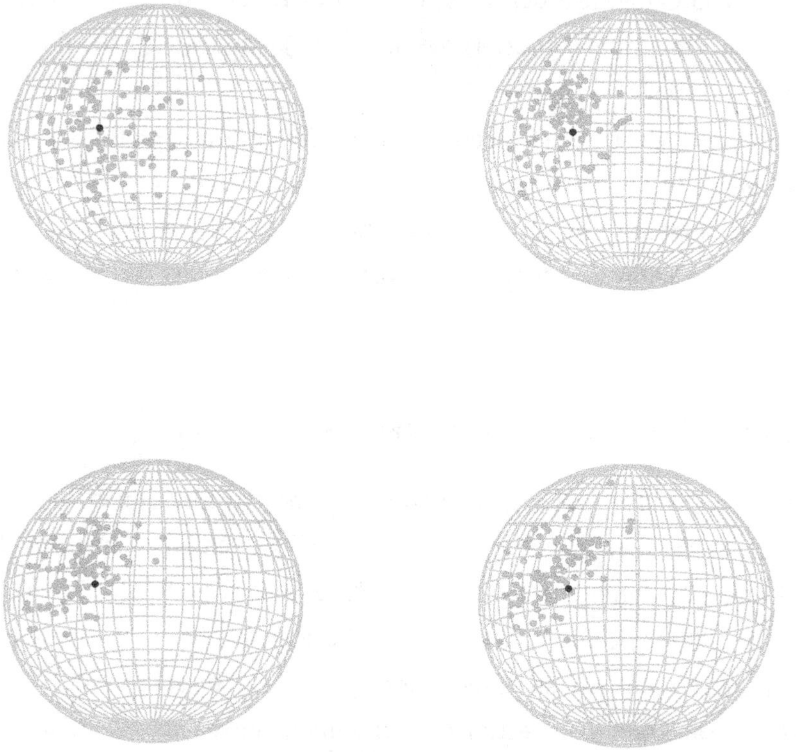

Figure 2.6: Illustration of skew-rotationally-symmetric distributions through the generation of 100 skew-FvML observations with concentration $\kappa = 10$. The data are increasingly skewed with (from top left to bottom right) $||\boldsymbol{\lambda}|| = 0, 2\sqrt{2}, 4\sqrt{2}$, and $8\sqrt{2}$. The black point indicates the location parameter $\boldsymbol{\mu}$ of the various distributions.

2.4 Flexible toroidal and cylindrical distributions

Modeling and analyzing data with two directional components or with a directional and a linear component has gained increasing interest in recent years, mainly motivated by datasets from emerging scientific disciplines such as bioinformatics and by advances in circular-circular and circular-linear regression.

In what follows, the vast majority of the directional components in toroidal and cylindrical models will be of a circular nature. Consequently, in most cases densities on the unit torus $\mathcal{S}^1 \times \mathcal{S}^1$ will be of the form $(\theta_1, \theta_2) \mapsto f_T(\theta_1, \theta_2)$ for $\theta_1, \theta_2 \in \mathcal{S}^1$ and cylindrical densities will be of the form $(\theta, z) \mapsto f_C(\theta, z)$ for $\theta \in \mathcal{S}^1$ and z lying on (a subset of) the real line; f_T and f_C are the focus of our interest.

The structure of the present section slightly differs from Sections 2.2 and 2.3 in that we imbed the classical models in the description of the recent flexible models. Since toroidal and cylindrical models have received less attention in earlier monographs, we begin with a historical note and motivations for the use of these models.

2.4.1 Some history, motivations and goals

The first proposals for probability distributions on the torus or cylinder can be traced back to the 1970s, under the impetus of the seminal papers by Mardia (1975), Mardia & Sutton (1978) and Johnson & Wehrly (1978). Between 1978 and the year 2000 there was almost a lull in the search for new distributions. This fact is reflected by only two short sections (3.7 and 11.4) on such models in Mardia & Jupp (2000) and one short section (2.3) in Jammalamadaka & SenGupta (2001), in contrast to both books dedicating an entire chapter (Chapter 11 in Mardia & Jupp 2000 and Chapter 8 in Jammalamadaka & SenGupta 2001) to directional-directional, respectively directional-linear, regression and correlation. We refer to Kato et al. (2008), Bhattacharya & SenGupta (2009) and Chapter 8 of Pewsey et al. (2013) for more recent information on those topics.

A strong resurgence of interest in toroidal and cylindrical models has occurred in recent years, motivated primarily by demands from other scientific domains. The main motivation for new toroidal distributions comes from structural bioinformatics where it is essential to model the joint distribution of dihedral angles when analyzing protein structures, see Section 1.2.5. Environmental sciences drive the research for flexible cylindrical models, especially the combination of wind direction with wind speed, temperature, SO_2 concentration or other air quality indicators. A more detailed example on wildfires is provided in Section 1.2.4. The study of animal behavior and oceanography are also driving forces behind advances in linear-circular analyses.

In addition to the desirable features identified in Section 2.2.3, one should also expect appealing toroidal and cylindrical models to possess

- tractable marginal and conditional distributions, ideally of well-known forms;
- a versatile dependence structure.

The former point is essential for regression purposes; the latter allows, for instance, modeling situations where the directional concentration increases with the length of the linear component.

The following sections are intended to draw the most complete picture of up-to-date advances in toroidal and cylindrical modeling.

2.4.2 The bivariate von Mises distribution and its variants

Mardia (1975) introduced the *bivariate von Mises (bvM) density* on the torus

$$\begin{aligned}(\theta_1, \theta_2) \mapsto C \exp\big(&\kappa_1 \cos(\theta_1 - \mu_1) + \kappa_2 \cos(\theta_2 - \mu_2) \\ &+ (\cos(\theta_1 - \mu_1), \sin(\theta_1 - \mu_1))\mathbf{A}(\cos(\theta_2 - \mu_2), \sin(\theta_2 - \mu_2))'\big), \end{aligned} \tag{2.24}$$

with normalizing constant C, circular location parameters $\mu_1, \mu_2 \in [-\pi, \pi)$, concentration parameters $\kappa_1, \kappa_2 \geq 0$ and circular-circular dependence parameter $\mathbf{A}$, a 2×2 matrix. The bvM distribution has the attractive property of being a maximum entropy distribution subject to constraints involving $\mathrm{E}[\cos(\Theta_1)]$, $\mathrm{E}[\sin(\Theta_1)]$, $\mathrm{E}[\cos(\Theta_2)]$, $\mathrm{E}[\sin(\Theta_2)]$, $\mathrm{E}[\cos(\Theta_1)\cos(\Theta_2)]$, $\mathrm{E}[\cos(\Theta_1)\sin(\Theta_2)]$, $\mathrm{E}[\sin(\Theta_1)\cos(\Theta_2)]$ and $\mathrm{E}[\sin(\Theta_1)\sin(\Theta_2)]$. With a total of eight parameters, the bvM is overparametrized, which is why special cases were proposed almost immediately after its introduction. In particular, Rivest (1988) considered the subclass of densities proportional to

$$\begin{aligned}\exp\big(&\kappa_1 \cos(\theta_1 - \mu_1) + \kappa_2 \cos(\theta_2 - \mu_2) + \alpha \cos(\theta_1 - \mu_1)\cos(\theta_2 - \mu_2) \\ &+ \beta \sin(\theta_1 - \mu_1)\sin(\theta_2 - \mu_2)\big), \end{aligned} \tag{2.25}$$

with $\alpha, \beta \in \mathbb{R}$. To achieve further parameter parsimony, Singh et al. (2002) particularized (2.25) to the setting where $\alpha = 0$, leading to the *Sine model* with density

$$(\theta_1, \theta_2) \mapsto C \exp\left(\kappa_1 \cos(\theta_1 - \mu_1) + \kappa_2 \cos(\theta_2 - \mu_2) + \beta \sin(\theta_1 - \mu_1)\sin(\theta_2 - \mu_2)\right)$$

where the normalizing constant C is given by

$$C^{-1} = 4\pi^2 \sum_{i=0}^{\infty} \binom{2i}{i} \left(\frac{\beta^2}{4\kappa_1\kappa_2}\right)^i I_i(\kappa_1) I_i(\kappa_2).$$

In the same spirit, Mardia et al. (2007) investigated (2.25) with $\alpha = \beta = -\kappa_3$, leading to the *Cosine model* with density

$$(\theta_1, \theta_2) \mapsto C \exp\left(\kappa_1 \cos(\theta_1 - \mu_1) + \kappa_2 \cos(\theta_2 - \mu_2) - \kappa_3 \cos(\theta_1 - \mu_1 - \theta_2 + \mu_2)\right),$$

where the normalizing constant C is given by

$$C^{-1} = 4\pi^2 \left[I_0(\kappa_1) I_0(\kappa_2) I_0(\kappa_3) + 2 \sum_{i=1}^{\infty} I_i(\kappa_1) I_i(\kappa_2) I_i(\kappa_3) \right].$$

The shapes of the Rivest, Sine and Cosine submodels, together with the independence model (with $\mathbf{A}$ being the zero matrix), are shown in Figure 2.7 in the form of contour plots. Yet another special case, which is intimately related to the Rivest (1988) model and reduces to the latter in certain circumstances, is the so-called hybrid model briefly described in Kent et al. (2008). An insightful review of the various variants of bivariate von Mises distributions is presented in Chapter 6 of Hamelryck et al. (2012).

The Sine and Cosine models have conditional von Mises densities, while the marginal densities are proportional to expressions of the form $I_0(h(\theta - \mu)) \exp(\kappa \cos(\theta - \mu))$ for some function h, some location $\mu \in [-\pi, \pi)$ and some concentration $\kappa \geq 0$. These marginal densities are symmetric but can be both unimodal and bimodal, depending on conditions involving the non-location parameters. The conditions for uni- or bimodality of the joint distributions are of a simpler form. We refer the reader to Mardia et al. (2007) for details, as well as for a thorough comparison of the Sine and Cosine models in their Section 3.2.

A multivariate extension of the Sine model with p angular components was proposed in Mardia et al. (2008). The density reads

$$(\theta_1, \ldots, \theta_p) \mapsto C_p \exp\left(\boldsymbol{\kappa}' c(\boldsymbol{\theta}, \boldsymbol{\mu}) + s(\boldsymbol{\theta}, \boldsymbol{\mu})' \boldsymbol{\beta} s(\boldsymbol{\theta}, \boldsymbol{\mu})\right),$$

where $c(\boldsymbol{\theta}, \boldsymbol{\mu}) = (\cos(\theta_1 - \mu_1), \ldots, \cos(\theta_p - \mu_p))'$, $s(\boldsymbol{\theta}, \boldsymbol{\mu}) = (\sin(\theta_1 - \mu_1), \ldots, \sin(\theta_p - \mu_p))'$, $\boldsymbol{\beta}$ is a real-valued symmetric matrix with zeroes on the

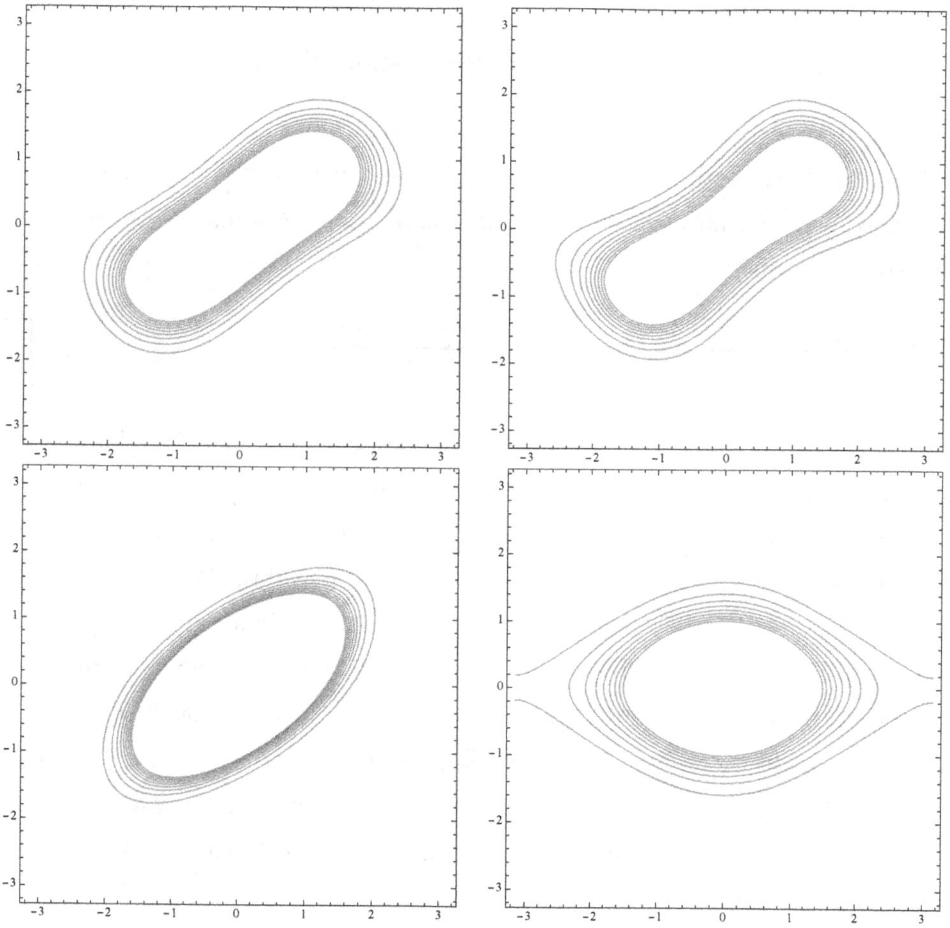

Figure 2.7: Contour plots of variants of the bivariate von Mises density with $a_{11} = 1, a_{12} = 0, a_{21} = 0, a_{22} = 4$, corresponding to the Rivest submodel (top left), $a_{11} = 0, a_{12} = 0, a_{21} = 0, a_{22} = 4$, corresponding to the Sine model (top right), $a_{11} = 4, a_{12} = 0, a_{21} = 0, a_{22} = 4$, corresponding to the Cosine model (bottom left), and $a_{11} = a_{12} = a_{21} = a_{22} = 0$, yielding the independence model (bottom right). In each case, $\mu_1 = \mu_2 = 0$ and the concentrations are $\kappa_1 = 2$ and $\kappa_2 = 4$.

diagonal, and C_p is a normalizing constant whose explicit form is unknown for $p > 2$. Similarly, a multivariate extension of the Cosine distribution was defined in Mardia & Patrangenaru (2005).

2.4.3 Mardia–Sutton type cylindrical distributions

Starting from the idea that a cylindrical model should have a simple dependence structure and that, in the case of independence, the linear and circular parts should respectively be the normal and von Mises distributions, Mardia & Sutton (1978) proposed the density

$$(\theta, z) \mapsto \frac{1}{\sigma(2\pi)^{3/2} I_0(\kappa)} \exp\left\{ -\frac{(z - (\mu' + \lambda \cos(\theta - \nu)))^2}{2\sigma^2} + \kappa \cos(\theta - \mu) \right\} \tag{2.26}$$

with circular location $\mu \in [-\pi, \pi)$ and concentration $\kappa \geq 0$, linear location $\mu' \in \mathbb{R}$ and dispersion $\sigma > 0$, and circular-linear parameters $\nu \in [-\pi, \pi)$ and $\lambda \geq 0$. The latter parameter regulates the dependence structure, $\lambda = 0$ corresponding to the product of independent von Mises and normal densities. We call (2.26) the *Mardia–Sutton density*. This distribution can be derived in a rather elegant way. Recalling that the von Mises distribution is obtained by conditioning on a bivariate normal distribution, the Mardia–Sutton density follows from a conditioning argument on a trivariate normal distribution whose first component is the linear part in (2.26). The highly desirable conditional distributions pave the way to regression analysis. The marginal circular density is again a von Mises, while the form of the marginal linear part is complicated.

Kato & Shimizu (2008) investigated a flexible extension of the Mardia–Sutton model, again by conditioning a trivariate normal distribution. The corresponding density is

$$(\theta, z) \mapsto C \exp\left(-\frac{(z - (\mu' + \lambda \cos(\theta - \nu)))^2}{2\sigma^2} + \kappa_1 \cos(\theta - \mu_1) + \kappa_2 \cos(2(\theta - \mu_2)) \right)$$

with $\sigma > 0$, $\kappa_1, \kappa_2 \geq 0$, $-\pi \leq \mu_1, \nu < \pi$, $0 \leq \mu_2 < \pi$, $\mu' \in \mathbb{R}$, $\lambda \geq 0$ and normalizing constant

$$C^{-1} = (2\pi)^{3/2} \sigma \left\{ I_0(\kappa_1) I_0(\kappa_2) + 2 \sum_{i=1}^{\infty} I_i(\kappa_2) I_{2i}(\kappa_1) \cos\{2i(\mu_1 - \mu_2)\} \right\}.$$

The Mardia–Sutton density is retrieved when $\kappa_2 = 0$. Contour plots of the Mardia–Sutton and Kato–Shimizu models are provided in Figure 2.8. As can be seen, the Kato–Shimizu density can have more than one mode. This has a simple reason:

the von Mises part in the Mardia–Sutton model is replaced with a Generalized von Mises, capable of modeling symmetry and asymmetry as well as unimodality and bimodality (see Section 2.2.8). Consequently, the circular conditional and marginal laws are GvM distributions, while the linear conditional distribution is a normal, as in the Mardia–Sutton model. Both models enjoy a maximum entropy characterization like the bivariate von Mises distribution of Section 2.4.2. Maximum likelihood estimation is straightforward, especially for the Mardia–Sutton model where the estimators have closed-form expressions.

Very recently, Sugasawa et al. (2015) proposed a new flexible cylindrical model obtained by conditioning a trivariate t distribution. They term it the generalized t-distribution on the cylinder.

2.4.4 Johnson–Wehrly type cylindrical distributions

Aiming to construct maximum entropy distributions under various moment conditions, Johnson & Wehrly (1978) proposed three different circular-linear densities plus an additional fourth density for higher-dimensional linear and directional vectors. Their first density is of the very simple form

$$(\theta, z) \mapsto \frac{(\lambda^2 - \kappa^2)^{1/2}}{2\pi} \exp\left(-\lambda z + \kappa z \cos(\theta - \mu)\right), \tag{2.27}$$

with circular location $\mu \in [-\pi, \pi)$, linear dispersion $\lambda > 0$ and circular-linear parameter $0 \leq \kappa < \lambda$ which governs dependence. The conditional linear and circular distributions are exponential and von Mises, respectively, while the marginal circular law is wrapped Cauchy. The marginal linear density takes the form $(\lambda^2 - \kappa^2) I_0(\kappa z) e^{-\lambda z}$.

The second Johnson–Wehrly density resembles the Mardia–Sutton model of Section 2.4.3:

$$(\theta, z) \mapsto C \frac{e^{-\kappa^2/(4\sigma^2)}}{\sqrt{2\pi}\sigma} \exp\left(-\frac{(z-\mu')^2}{2\sigma^2} + \frac{\kappa z}{\sigma^2}\cos(\theta - \mu)\right), \tag{2.28}$$

with

$$C^{-1} = 2\pi \left(I_0\left(\frac{\kappa\mu'}{\sigma^2}\right) I_0\left(\frac{\kappa^2}{4\sigma^2}\right) + 2\sum_{i=1}^{\infty} I_i\left(\frac{\kappa^2}{4\sigma^2}\right) I_{2i}\left(\frac{\kappa\mu'}{\sigma^2}\right)\right)$$

and circular location $\mu \in [-\pi, \pi)$, linear location $\mu' \in \mathbb{R}$ and dispersion $\sigma > 0$, and dependence parameter $\kappa \geq 0$. The conditional laws are von Mises and normal,

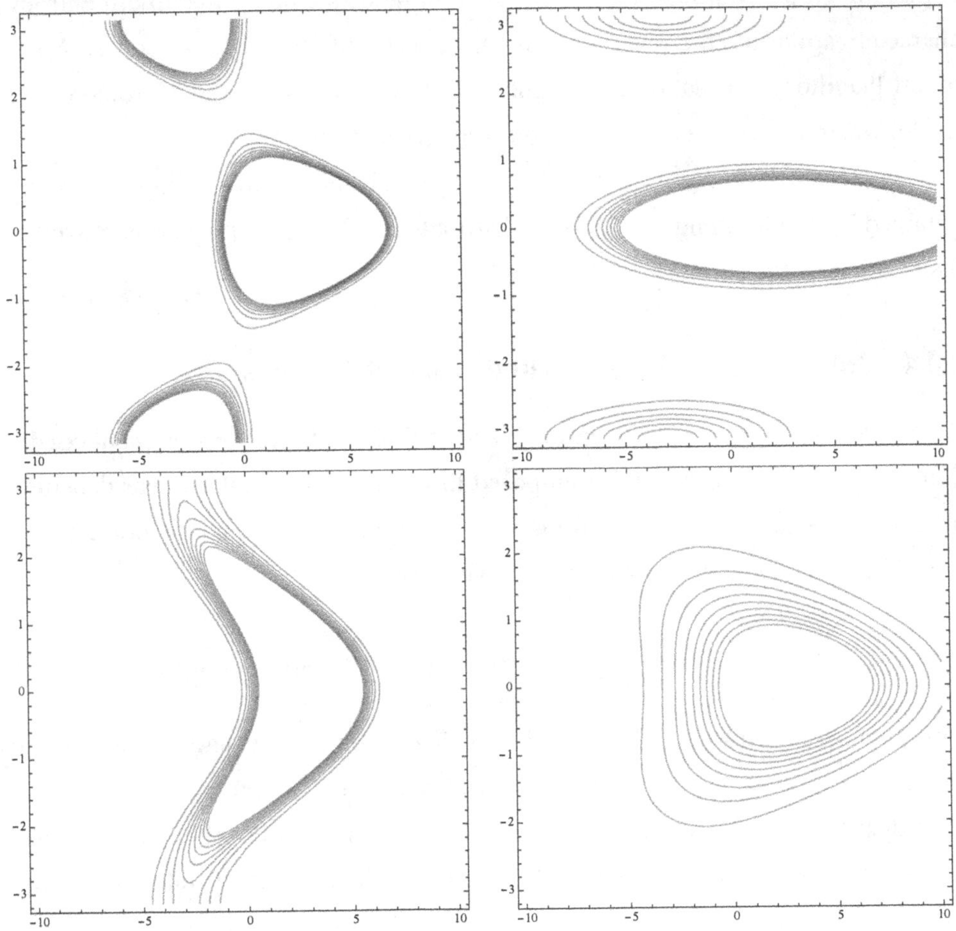

Figure 2.8: Contour plots of the Kato–Shimizu (top) and Mardia–Sutton (bottom) cylindrical densities with $\kappa_1 = 2, \kappa_2 = 4, \sigma = 1, \lambda = 3$ (top left), $\kappa_1 = 2, \kappa_2 = 4, \sigma = 3, \lambda = 3$ (top right), $\kappa_1 = 2, \kappa_2 = 0, \sigma = 1, \lambda = 3$ (bottom left), and $\kappa_1 = 2, \kappa_2 = 0, \sigma = 3, \lambda = 3$ (bottom right). In each case, $\mu_1 = \mu_2 = 0$.

as for (2.26), the marginal circular is GvM while the marginal linear is proportional to $\exp(-(z-\mu')^2/(2\sigma^2))I_0(\kappa z/\sigma^2)$.

A common drawback of the first two Johnson–Wehrly distributions (2.27, 2.28) lies in the dependence structure: the circular part becomes uniform when $\kappa = 0$, i.e., in the case of independence. In order to overcome this limitation, Johnson and Wehrly proposed a third model of the form

$$(\theta, z) \mapsto \frac{(\lambda^2-\kappa^2)^{1/2}}{2\pi\left(I_0(\nu)+2\sum_{i=1}^{\infty}\rho^i I_i(\nu)\cos(i(\mu_1-\mu_2))\right)} \times \exp\left(-\lambda z+\kappa z\cos(\theta-\mu_1)+\nu\cos(\theta-\mu_2)\right),$$

with circular locations $\mu_1, \mu_2 \in [-\pi, \pi)$, circular concentration $\nu \geq 0$, linear dispersion $\lambda > 0$ and dependence parameter $0 \leq \kappa < \lambda$, and where $\rho = \kappa(\lambda + (\lambda^2-\kappa^2)^{1/2})^{-1}$. As for (2.27), the conditional laws are von Mises and exponential, but here the marginal laws are less tractable. When $\kappa = 0$, the circular part becomes von Mises, as was the motivation underpinning this third model. A drawback is the normalizing constant which involves an infinite sum.

Abe & Ley (2017) introduced a Johnson–Wehrly-like model that overcomes the shortcomings of (2.27) and (2.29). Their density reads

$$(\theta, z) \mapsto \frac{\alpha\beta^\alpha}{2\pi\cosh(\kappa)}\left(1+\lambda\sin(\theta-\mu)\right) z^{\alpha-1}\exp\left(-(\beta z)^\alpha\left(1-\tanh(\kappa)\cos(\theta-\mu)\right)\right) \tag{2.29}$$

with circular location $\mu \in [-\pi, \pi)$ and skewness $\lambda \in (-1, 1)$, linear dispersion $\beta > 0$ and shape parameter $\alpha > 0$, and where $\kappa \geq 0$ is both a circular concentration and the cylindrical dependence parameter, as in the other Johnson–Wehrly distributions. The multiplication by $(1 + \lambda\sin(\theta - \mu))$ is reminiscent of the sine-skewing in Section 2.2.5. Indeed, the Abe–Ley density increases the flexibility of (2.27) in several ways: the linear exponential part is turned into the more general Weibull distribution, the circular von Mises part becomes sine-skewed, and the latter transformation also entails that, when $\kappa = 0$, the circular part of (2.29) is cardioid instead of uniform. Figure 2.9 exhibits contour plots of the Abe–Ley and first Johnson–Wehrly densities for various parameter choices.

The Johnson–Wehrly model (2.27) is retrieved when $\lambda = 0$ and $\alpha = 1$. Compared to (2.29), the Abe–Ley density enjoys the attractive feature of a very simple normalizing constant, rendering, for instance, moment calculations simple. The conditional circular and linear laws are sine-skewed von Mises and Weibull, respectively, while the circular marginal law is sine-skewed wrapped Cauchy. The linear marginal density is proportional to

$I_0(z^\alpha \beta^\alpha \tanh(\kappa)) z^{\alpha-1} \exp(-(\beta z)^\alpha)$. Circular-linear regression is straightforward, and parameter estimation works well via maximum likelihood. A close inspection of the conditional circular law reveals that the concentration parameter of the sine-skewed von Mises is $(\beta z)^\alpha \tanh(\kappa)$, showing that the Abe–Ley model as well as its Johnson–Wehrly submodel are able to model cylindrical data where the circular concentration increases with the linear part (see Section 2.4.1).

2.4.5 The copula approach

A popular means of constructing flexible distributions in $\mathbb{R}^p$ is provided by copulas, as they accommodate modeling the marginal distributions and the dependence structure between the p components separately; see Joe (1997) or Nelsen (2010) for monographs on copulas. It is thus hardly surprising that copula-like structures have been put forward for toroidal and cylindrical distributions, too. Under their most general form, copula-based densities read

$$(\theta, y) \mapsto c\,(F_1(\theta), F_2(y))\, f_1(\theta) f_2(y) \tag{2.30}$$

where y is either a linear or a circular component, f_1 and F_1 (respectively, f_2 and F_2) are a circular (respectively, a linear or a circular) density and its associated distribution function, and c is a bivariate copula density regulating the dependence. In the toroidal setting, c is a density on the torus with uniform marginals. Jones et al. (2015) coin the term *circula* for the copula c in such settings. Densities (2.30) have not yet been studied in full generality and will certainly, in the near future, lead to interesting new cylindrical and toroidal probability laws with specified marginals f_1 and f_2 (or with p marginals, in the p-dimensional setting).

One particular case of (2.30) has received considerable attention in the literature. It corresponds to

$$(\theta, y) \mapsto 2\pi c_b\big(2\pi\,(F_1(\theta) - qF_2(y))\big) f_1(\theta) f_2(y), \tag{2.31}$$

with $q = 1$ in the cylindrical and $q = 1$ or -1 in the toroidal setting, and with c_b a circular density called binding function. This choice was popularized in Johnson & Wehrly (1978) and Wehrly & Johnson (1980), with no mention of copulas. Johnson & Wehrly (1977) and Isham (1977) had previously used similar constructions. An in-depth study of the copula-structure (2.31) is provided in Jones et al. (2015), and parametric bootstrap goodness-of-fit tests for the toroidal case are given in Pewsey & Kato (2016).

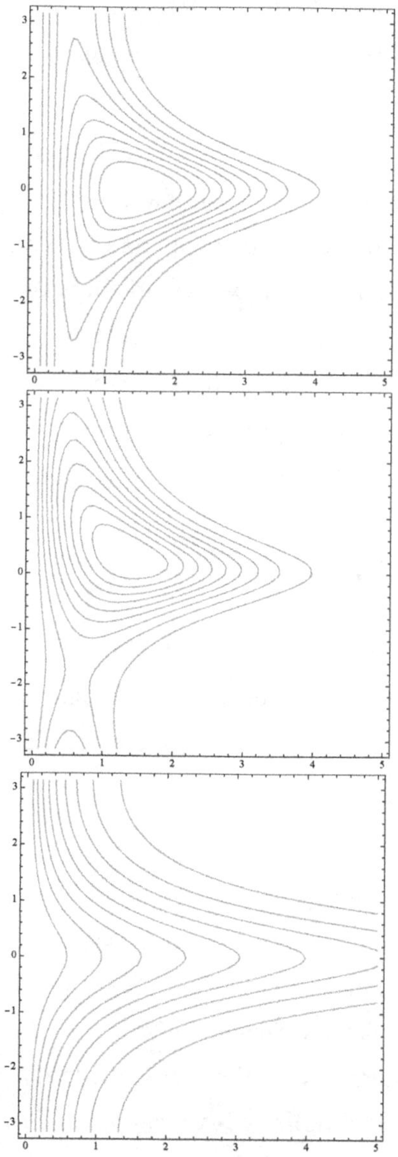

Figure 2.9: Contour plot of the Abe–Ley cylindrical density with $\alpha = 2, \lambda = 0$ (top), $\alpha = 2, \lambda = 0.5$ (middle) and $\alpha = 1, \lambda = 0$, which corresponds to the first Johnson–Wehrly density (bottom). In each case, μ is fixed to 0, β to 1 and κ to 1.

One special case of (2.31) is particularly interesting on the torus, namely the bivariate wrapped Cauchy (bwC) model recently proposed in Kato & Pewsey (2015). Choosing both marginals as well as the binding function to be wrapped Cauchy leads to a density proportional to the inverse of

$$(c_0-c_1\cos(\theta_1-\mu_1)-c_2\cos(\theta_2-\mu_2)-c_3\cos(\theta_1-\mu_1)\cos(\theta_2-\mu_2)-c_4\sin(\theta_1-\mu_1)\sin(\theta_2-\mu_2))$$

for constants c_1, c_2, c_3, c_4 depending on two circular concentration parameters and a dependence parameter. The bwC enjoys several nice properties such as unimodality, clear parameter interpretability, a simple normalizing constant, closed-form expressions for the trigonometric moments and hence fast method of moments estimation of its parameters and, most remarkably, also the circular conditional distributions are wrapped Cauchy.

Numerous further copula-based models have been investigated in the literature, and we refer to Jones et al. (2015) for detailed information and to the part "On the torus and the cylinder" of the next section for references.

2.5 Further reading

So far in this chapter we have attempted to describe the main recent advances in the flexible modeling of directional data, with particular focus on the circular setting. We now provide further references for the interested reader.

On the circle

We have briefly presented the classical distributions such as the von Mises, cardioid and wrapped Cauchy. For more details on these aspects and other classical circular distributions, we refer the interested reader to Section 3.5 of Mardia & Jupp (2000), Chapter 2 of Jammalamadaka & SenGupta (2001) and Section 4.3 of Pewsey et al. (2013). Of particular interest is Section 3.5.1 of Mardia & Jupp (2000) where the authors describe the broad families of *exponential* and *transformation* models and their respective links with inferential procedures. This source nicely complements the present chapter, where we have purposely not referred to such issues.

As examples of wrapped distributions, we have considered the wrapped normal and the wrapped Cauchy. However, new wrapped circular models have seen the light over the last two decades. Unimodal symmetric proposals include the wrapped t densities of Pewsey et al. (2007), while unimodal asymmetric models are the wrapped skew-normal (Pewsey 2000), the wrapped (skew) Laplace (Jammalamadaka & Kozubowski 2003, Jammalamadaka & Kozubowski 2004), the wrapped normal-Laplace and wrapped generalized normal-Laplace (Reed & Pewsey 2009). The unimodal wrapped stable distribution, first appearing in Mardia (1972), has received particular attention: in its three-parameter symmetric form by Jammalamadaka & SenGupta (2001), Section 2.2.8, SenGupta & Pal (2001) and Gatto & Jammalamadaka (2003), and in its more flexible four-parameter, potentially skew, form by Pewsey (2008).

Besides wrapping, inverse stereographic projection via the tangent function is another approach which can be used to convert linear densities into circular ones. Minh & Farnum (2003) applied this method to obtain a circular t distribution which happens to coincide with Cartwright's power-of-cosine distribution (see Section 2.2.4), while Abe et al. (2010) combined this procedure with the structure inherent to the Jones–Pewsey family to obtain a new flexible symmetric unimodal distribution.

We have mentioned on various occasions finite mixture distributions as a means of producing multimodal distributions. The best known and most thoroughly studied mixtures to date are mixtures of von Mises distributions (Mardia & Sutton 1975, Spurr & Koutbeiy 1991), which have been reconsidered in, e.g., Mooney et al. (2003). See also Section 5.5 of Mardia & Jupp (2000), Section 4.3 of Jammalamadaka & SenGupta (2001) and Section 4.3.16 of Pewsey et al. (2013).

An approach similar in spirit to finite mixtures was adopted by Fernández-Durán (2004) who considered non-negative trigonometric sums. The resulting densities are $\theta \mapsto \frac{1}{2\pi} + \frac{1}{\pi}\sum_{j=1}^{n}(a_j\cos(j\theta) + b_j\sin(j\theta))$ with $a_j - ib_j = 2\sum_{\nu=0}^{n-j} c_{\nu+j}\bar{c}_\nu$ for complex numbers c_j (with $\bar{c}_j$ the complex conjugate of c_j) such that $\sum_{j=0}^{n}|c_j|^2 = (2\pi)^{-1}$. The number n of terms in the sum is not fixed and can become unpleasantly large for certain data shapes, hence is an additional parameter. In particular, $n = 0$ produces the uniform and $n = 1$ the cardioid. For inferential issues for such densities we refer the reader to Fernández-Durán & Gregorio-Domínguez (2010).

Throughout this chapter we have already mentioned the t distribution on the circle proposed by Shimizu & Iida (2002), which belongs to the Jones–Pewsey family, as well as the wrapped t distribution of Pewsey et al. (2007). Yet another circular t distribution, the generalized t distribution, was introduced in Siew et al. (2008).

On the (hyper-)sphere

As for the circular case, multimodal spherical data are in general best modeled using finite mixtures of popular unimodal distributions. Multimodal spherical datasets appear in a variety of disciplines such as the study of fractures in rock masses (Peel et al. 2001), genetics and text mining (Banerjee et al. 2005). Notable recent contributions in this direction are finite mixtures of Kent distributions, studied in Peel et al. (2001), and mixtures of FvML distributions (Banerjee et al. 2005, Ferreira et al. 2008).

Finite mixtures of FvML distributions were actually not the main goal of Ferreira et al. (2008). These authors constructed directional distributions able to cope with both skewness and multimodality, which they called *directional log-spline distributions*. Their densities are expressed as splines on hyperspheres (Taijeron et al. 1994), which explains their flexibility because hyperspherical splines can approximate any continuous function.

For symmetric axial data, which can be seen as a special form of bimodality, Oualkacha & Rivest (2009) introduced a new distribution, which they applied to a human body data set.

On the torus and the cylinder

In Section 2.4.5 we have described a copula-like approach to modeling toroidal and cylindrical data. In addition to the sources mentioned in that section, several other authors have investigated particular instances of model (2.31). Shieh & Johnson (2005) studied a toroidal distribution with von Mises marginals where c_b is also a von Mises density. Fernández-Durán (2007) used the non-negative trigonometric sums circular densities (see "On the circle" above) for c_b as well as for both circular densities in the toroidal case and for the single circular density in the cylindrical setting (with linear Weibull density). Shieh et al. (2011) proposed a model with Generalized von Mises marginals and c_b of von Mises type.

A particularly notable contribution is Kato (2009). We refer to that paper for further insight into the copula-like structure, as well as for toroidal distributions with, on the one hand, von Mises marginals and, on the other, uniform marginals.

Finally, we wish to mention Jupp (2015) where copulae on a product of Riemannian manifolds are studied.

Advances in kernel density estimation on directional supports

3.1 Introduction

We described in Chapter 2 a plethora of directional densities designed to best describe the directional data at hand. That approach is fully parametric, as the only unknown parts of the density are the parameters which we need to estimate. A completely different path is taken in the present chapter, where no underlying parametric density is assumed and where, hence, the density itself has to be estimated.

3.1.1 Kernel density estimation on the real line

Kernel density estimation is the classical way to produce non-parametric density estimates on the real line. Its roots can be found in the seminal works of Rosenblatt (1956) and Parzen (1962). Letting $Z_1, \ldots, Z_n$ be iid observations from a population with unknown density f on $\mathbb{R}$, the kernel density estimator (KDE) at some point $z \in \mathbb{R}$ is defined as

$$\hat{f}_g(z) = \frac{1}{ng} \sum_{i=1}^{n} K_\ell \left(\frac{z - Z_i}{g} \right) \tag{3.1}$$

where the *linear kernel* K_ℓ is a non-negative, continuous, bounded and symmetric function integrating to one, and $g > 0$ is the *bandwidth* parameter controlling the smoothness of the estimator. Note that here the index ℓ in K_ℓ indicates a kernel designed for linear data, as opposed to K_s for spherical data, see (3.4). The conditions on the kernel ensure that $\hat{f}_g$ integrates to one. Plainly speaking, the KDE (3.1) averages, at each point z, the impact of the observations via the quantities $\frac{1}{g} K_\ell \left(\frac{z - Z_i}{g} \right)$, $i = 1, \ldots, n$, where the effect of each Z_i depends on its distance from z, re-scaled through division by g. The precise choice of K_ℓ has low impact on the overall shape of $\hat{f}_g$; usually the kernel is a fast decaying function taking its maximal value at 0. Typical examples are densities on $\mathbb{R}$, especially the standard normal density. In contrast, the choice of the bandwidth g is crucial: large values of g lead to oversmoothed, small values to undersmoothed, estimates $\hat{f}_g$. Choosing a good bandwidth is a challenging problem, and numerous papers have addressed the *bandwidth selection* issue. We refer the reader to Chiu (1996) for a review and references.

The difficulty of finding a reasonable bandwidth is well illustrated by the bias and variance expressions of $\hat{f}_g(z)$. Assuming that f is twice differentiable, that the quantities

$$\mu_2(K_\ell) = \int_{\mathbb{R}} z^2 K_\ell(z)dz$$

and $\int_{\mathbb{R}} K_\ell^2(z)dz$ are finite and that the bandwidth $g = g_n$ is a sequence of numbers such that $g \to 0$ and $ng \to \infty$ as $n \to \infty$, the bias of $\hat{f}_g(z)$ is given by

$$\mathrm{E}\left[\hat{f}_g(z)\right] - f(z) = \frac{1}{2}\mu_2(K_\ell)f''(z)g^2 + o(g^2) \tag{3.2}$$

while the variance is

$$\mathrm{Var}\left[\hat{f}_g(z)\right] = \frac{1}{ng}R(K_\ell)f(z) + o\left((ng)^{-1}\right) \tag{3.3}$$

with

$$R(\psi) := \int_{\mathbb{R}} \psi^2(z)dz$$

for some function ψ on $\mathbb{R}$. Expressions (3.2) and (3.3) show that the bias is of order $O(g^2)$ and the variance of order $O\left((ng)^{-1}\right)$, underlining the ambiguous role played by the bandwidth g and the need to find a compromise between quick and slow convergence to zero. Minimization of the mean squared error, MSE $=$ bias^2 $+$ variance, is achieved for g proportional to $n^{-1/5}$.

There are various ways of measuring the performance of a density estimate, based on distinct error criteria. The MSE is a local criterion. A commonly adopted global error measurement criterion is the *Mean Integrated Squared Error (MISE)*

$$\mathrm{MISE}\left[\hat{f}_g\right] = \mathrm{E}\left[\int_{\mathbb{R}} \left(\hat{f}_g(z) - f(z)\right)^2 dz\right].$$

The integral alone, without the expectation, also defines a criterion, namely the *Integrated Squared Error*. Under additional regularity conditions, the MISE can be expressed as

$$\mathrm{MISE}\left[\hat{f}_g\right] = \frac{1}{4}(\mu_2(K_\ell))^2 R(f'')g^4 + \frac{1}{ng}R(K_\ell) + o\left(g^4 + (ng)^{-1}\right)$$

where the part $\frac{1}{4}(\mu_2(K_\ell))^2 R(f'')g^4 + \frac{1}{ng}R(K_\ell)$ is referred to as the *Asymptotic MISE* (AMISE). Minimizing the AMISE in g is a simple procedure to find as optimal bandwidth

$$g_{\text{AMISE}} = \left[\frac{R(K_\ell)}{(\mu_2(K_\ell))^2 R(f'')n}\right]^{1/5}$$

which is of order $O\left(n^{-1/5}\right)$. To be of practical use, the *curvature* term $R(f'')$ involving the unknown f needs to be estimated. Silverman (1986) suggested the rule of thumb of calculating $R(f'')$ on basis of the normal distribution as reference density f.

We recommend the interested reader who wishes to learn more about kernel density estimation on $\mathbb{R}$ as well as on $\mathbb{R}^p$ to consult the books by Silverman (1986), Wand & Jones (1995) and Scott (2015).

3.1.2 Organization of the remainder of the chapter

The present chapter will take the reader on a journey through recent advances in non-parametric density estimation for directional data. The focus will lie on kernel estimation methods; a very successful alternative approach is based on spherical needlets (Baldi et al. 2009). After defining the main concepts of directional kernel density estimation in Section 3.2, we shall discuss in detail the important issue of bandwidth selection in Section 3.3 and finally describe inferential procedures based on kernel density estimation in Section 3.4.

3.2 Definitions and main properties

We shall not describe in detail kernel density estimation for all kinds of directional supports in order to avoid repetition. Besides kernel density estimation on spheres (and hence, circles), we have opted to present cylindrical kernel density estimation as this combines directional with linear variables. For toroidal kernel density estimation, we refer the reader to the papers by Di Marzio et al. (2011) and Taylor et al. (2012).

3.2.1 Spherical kernel density estimation

Kernel density estimates on (hyper-)spheres were proposed by Hall et al. (1987) and Bai et al. (1988).[1] For iid observations $\mathbf{X}_1, \ldots, \mathbf{X}_n \in \mathcal{S}^{p-1}$ from a population with unknown density f, the corresponding spherical KDE at some $\mathbf{x} \in \mathcal{S}^{p-1}$ is given by

$$\hat{f}_h(\mathbf{x}) = \frac{c_{h,p}(K_s)}{n} \sum_{i=1}^{n} K_s\left(\frac{1 - \mathbf{x}'\mathbf{X}_i}{h^2}\right) \tag{3.4}$$

with bandwidth parameter $h = h_n > 0$, spherical kernel K_s and normalizing constant $c_{h,p}(K_s)$ to be discussed below. There are a couple of differences between (3.4) and the linear KDE (3.1). The linear distance $z - Z_i$ is replaced with a natural angular distance[2] $1 - \mathbf{x}'\mathbf{X}_i$ varying between 0 when $\mathbf{X}_i = \mathbf{x}$ and 2 when $\mathbf{X}_i = -\mathbf{x}$. Consequently, the spherical kernel K_s is a non-negative function defined on $\mathbb{R}^+$ that must also satisfy $\int_{\mathcal{S}^{p-1}} K_s\left(\frac{1-\mathbf{x}'\mathbf{y}}{h^2}\right) d\sigma_{p-1}(\mathbf{y}) < \infty$ for all $\mathbf{x} \in \mathcal{S}^{p-1}$. Recalling the tangent-normal decomposition (2.21), the latter condition can be reexpressed in more familiar terms as

$$\int_{-1}^{1} K_s\left(\frac{1-t}{h^2}\right) (1-t^2)^{(p-3)/2} dt < \infty$$
$$\iff \int_{0}^{2/h^2} K_s(v) v^{(p-3)/2} (2 - vh^2)^{(p-3)/2} dv < \infty.$$

Since $h_n \to 0$ when $n \to \infty$, the latter expression can be shown to be finite by the dominated convergence theorem under the condition

$$\int_{0}^{\infty} K_s(v) v^{(p-3)/2} dv < \infty. \tag{3.5}$$

The spherical kernel has to satisfy this condition for every dimension $p \geq 2$, which means that K_s is rapidly decaying, more rapidly than any polynomial. A typical example is $K_s(v) = e^{-v}$, the von Mises or FvML kernel. This terminology follows from the fact that the resulting KDE can then be seen as a mixture of FvML densities where each FvML component has $\mathbf{X}_i$ as location and $1/h^2$ as concentration parameter. Contrary to the linear setting, the spherical kernel K_s is not a density,

[1] These papers are usually considered as initiators of research on spherical kernel density estimation. It is however worth mentioning that Beran (1979*a*) had previously analyzed in his Section 3.2 a spherical KDE, as a by-product of investigations on exponential models on the sphere. Neither Hall et al. (1987) nor Bai et al. (1988) mentioned that paper.

[2] It is interesting to note that $1 - \mathbf{x}'\mathbf{X}_i = \frac{1}{2}||\mathbf{x} - \mathbf{X}_i||^2$. The use of a squared distance between $\mathbf{x}$ and $\mathbf{X}_i$ provides an explanation as to why the bandwidth h is squared in (3.4).

which explains the need for the normalizing constant $c_{h,p}(K_s)$ in (3.4). The latter is given by

$$\begin{aligned} c_{h,p}(K_s)^{-1} &= \int_{\mathcal{S}^{p-1}} K_s\left(\frac{1-\mathbf{x}'\mathbf{y}}{h^2}\right) d\sigma_{p-1}(\mathbf{y}) \\ &= h^{p-1}\omega_{p-1}\int_0^{2/h^2} K_s(v)v^{(p-3)/2}(2-vh^2)^{(p-3)/2}dv \qquad (3.6) \\ &= O(h^{p-1}) \end{aligned}$$

as $n \to \infty$ under the condition specified above. The asymptotic $O(h^{p-1})$ behavior follows from the same developments that led to (3.5).

Conditions for pointwise, uniform and L_1-norm strong consistency are studied in Bai et al. (1988). Bias and variance formulae for the spherical KDE $\hat{f}_h(\mathbf{x})$ were investigated in Hall et al. (1987), Klemelä (2000) and Zhao & Wu (2001). The following conditions must be verified by the density f, the kernel K_s and the bandwidth h to derive these results:

- SKDE1: Extending f from $\mathcal{S}^{p-1}$ to $\mathbb{R}^p$ via $f(\mathbf{x}) \to f(\mathbf{z}/||\mathbf{z}||)$ for $\mathbf{z} \neq \mathbf{0}$, the gradient vector $\nabla f(\mathbf{z}) = \left(\frac{\partial f(\mathbf{z})}{\partial z_1}, \ldots, \frac{\partial f(\mathbf{z})}{\partial z_p}\right)'$ and the Hessian matrix $\mathbf{H}f(\mathbf{z}) = \left(\frac{\partial f(\mathbf{z})}{\partial z_i \partial z_j}\right)_{1\le i,j\le p}$ exist and are continuous on $\mathbb{R}^p\backslash\{\mathbf{0}\}$ (hence they are square-integrable on $\mathcal{S}^{p-1}$).

- SKDE2: The kernel K_s satisfies condition (3.5) and (only for the variance)

$$\int_0^\infty K_s^2(v)v^{(p-3)/2}dv < \infty.$$

- SKDE3: Assuming that $h = h_n$, the bandwidth is a sequence of positive numbers such that $h_n \to 0$ and $nh_n^{p-1} \to \infty$ as $n \to \infty$.

The expressions for the bias and variance at some point $\mathbf{x} \in \mathcal{S}^{p-1}$ are then given by

$$\mathrm{E}\left[\hat{f}_h(\mathbf{x})\right] - f(\mathbf{x}) = b_p(K_s)\Psi(f,\mathbf{x})h^2 + o(h^2) \qquad (3.7)$$

with

$$\begin{aligned} b_p(K_s) &= \frac{\int_0^\infty K_s(v)v^{(p-1)/2}dv}{\int_0^\infty K_s(v)v^{(p-3)/2}dv} \\ \Psi(f,\mathbf{x}) &= (p-1)^{-1}(\nabla^2 f(\mathbf{x}) - \mathbf{x}'\mathbf{H}f(\mathbf{x})\mathbf{x}), \qquad (3.8) \end{aligned}$$

and

$$\mathrm{Var}\left[\hat{f}_h(\mathbf{x})\right] = \frac{c_{h,p}(K_s)}{n}d_p(K_s)f(\mathbf{x}) + o\left((nh^{p-1})^{-1}\right) \qquad (3.9)$$

with

$$d_p(K_s) = \frac{\int_0^\infty K_s^2(v)v^{(p-3)/2}dv}{\int_0^\infty K_s(v)v^{(p-3)/2}dv},$$

respectively. The notation $\nabla^2 f(\mathbf{x})$ in (3.8) stands for the Laplacian $\sum_{i=1}^p \frac{\partial^2 f(\mathbf{x})}{\partial x_i^2}$. The bias is of order $O(h^2)$ and involves the curvature of f via the Hessian matrix. From (3.6) it is readily seen that the variance is of order $O\left((nh^{p-1})^{-1}\right)$. We attract the reader's attention to the potentially misleading fact that often in the literature the quantity $-\mathbf{x}'\boldsymbol{\nabla} f(\mathbf{x})$ is present in the expression of $\boldsymbol{\Psi}(f, \mathbf{x})$. This is not an error but just not needed, because $\mathbf{x}'\boldsymbol{\nabla} f(\mathbf{x}) = 0$ as can be seen by extending $\mathbf{x} \in S^{p-1}$ into $\mathbf{x}/||\mathbf{x}||$ and then noting that $\boldsymbol{\nabla} f(\mathbf{x}) = (\mathbf{I}_p - \mathbf{x}\mathbf{x}')\boldsymbol{\nabla} f(\mathbf{x})$.

3.2.2 Cylindrical kernel density estimation

Kernel density estimation in the spherical-linear setting was first considered in García-Portugués et al. (2013*b*). Considering iid observations $(\mathbf{X}_1, Z_1), \ldots, (\mathbf{X}_n, Z_n) \in S^{p-1} \times \mathbb{R}$ from a population with unknown density $f(\mathbf{x}, z)$, the spherical-linear or cylindrical kernel density estimate is defined as

$$\hat{f}_{h,g}(\mathbf{x}, z) = \frac{c_{h,p}(K_s)}{ng} \sum_{i=1}^n K_s\left(\frac{1 - \mathbf{x}'\mathbf{X}_i}{h^2}\right) K_\ell\left(\frac{z - Z_i}{g}\right) \tag{3.10}$$

with bandwidth parameters $h, g > 0$ and normalizing constant $c_{h,p}(K_s)$ from (3.6). The kernel product structure $K_s(\cdot)K_\ell(\cdot)$ could be replaced with a general cylindrical kernel $K_c\left(\frac{1-\mathbf{x}'\mathbf{X}_i}{h^2}, \frac{z-Z_i}{g}\right)$ that links the spherical and linear parts in a more complicated way.

The gradient vector $\boldsymbol{\nabla} f(\mathbf{x}, z) = ((\boldsymbol{\nabla}_\mathbf{x} f(\mathbf{x}, z))', \nabla_z f(\mathbf{x}, z))'$ and Hessian matrix

$$\mathbf{H} f(\mathbf{x}, z) = \begin{pmatrix} \mathbf{H}_\mathbf{x} f(\mathbf{x}, z) & \mathbf{H}_{\mathbf{x},z} f(\mathbf{x}, z) \\ \mathbf{H}'_{\mathbf{x},z} f(\mathbf{x}, z) & H_z f(\mathbf{x}, z) \end{pmatrix}$$

are straightforward extensions of $\boldsymbol{\nabla} f(\mathbf{x})$ and $\mathbf{H} f(\mathbf{x})$ from the spherical setting (see Section 3.2.1). Adapting conditions SKDE1-SKDE2 to the cylindrical setting is straightforward, while the bandwidth parameters $h = h_n$ and $g = g_n$ need to satisfy $h_n \to 0, g_n \to 0$ and $nh_n^{p-1} g_n \to \infty$ as $n \to \infty$. The bias and variance expressions at some point $(\mathbf{x}, z) \in S^{p-1} \times \mathbb{R}$ then correspond to

$$\mathrm{E}\left[\hat{f}_{h,g}(\mathbf{x}, z)\right] - f(\mathbf{x}, z) = b_p(K_s)\boldsymbol{\Psi}_\mathbf{x}(f, \mathbf{x}, z)h^2 + \frac{1}{2}\mu_2(K_\ell)H_z f(\mathbf{x}, z)g^2 + o(h^2 + g^2) \tag{3.11}$$

with

$$\Psi_{\mathbf{x}}(f,\mathbf{x},z) = (p-1)^{-1}\left(\nabla^2_{\mathbf{x}} f(\mathbf{x},z) - \mathbf{x}'\mathbf{H}_{\mathbf{x}} f(\mathbf{x},z)\mathbf{x}\right)$$

where

$$\nabla^2_{\mathbf{x}} f(\mathbf{x},z) = \sum_{i=1}^{p} \frac{\partial^2 f(\mathbf{x},z)}{\partial x_i^2},$$

and

$$\mathrm{Var}\left[\hat{f}_{h,g}(\mathbf{x},z)\right] = \frac{c_{h,p}(K_s)}{ng} d_p(K_s) R(K_\ell) f(\mathbf{x},z) + o\left((nh^{p-1}g)^{-1}\right). \quad (3.12)$$

Pleasingly, the spherical and linear parts are identified in both the bias (under addition form) and the variance (under product form). This is also reflected in the respective orders, the bias having order $O(h^2+g^2)$ and the variance $O\left((nh^{p-1}g)^{-1}\right)$. Asymptotic normality of $\hat{f}_{h,g}(\mathbf{x},z)$ holds under an additional smoothness assumption on the product kernel $K_s(\cdot)K_\ell(\cdot)$; see García-Portugués et al. (2013*b*).

3.3 A delicate yet crucial issue: bandwidth choice

As in the linear setting, the choice of the bandwidth plays a crucial role in directional kernel density estimation. Figure 3.1 provides the reader with an idea of the impact of the bandwidth choice for circular data, hereby using three bandwidth selectors that are described in the next sections.

3.3.1 Spherical AMISE and bandwidth selection

As in the linear setting, minimization of the Asymptotic Mean Integrated Squared Error is a natural way to tackle the bandwidth selection problem. The spherical MISE, and hence the AMISE, is readily obtained from the bias and variance expressions from Section 3.2.1. Indeed, (3.7) and (3.9) lead to

$$\mathrm{AMISE}\left[\hat{f}_h(\mathbf{x})\right] = (b_p(K_s))^2 R(\Psi,f)h^4 + \frac{c_{h,p}(K_s)}{n} d_p(K_s), \quad (3.13)$$

with curvature term $R(\Psi,f) = \int_{\mathcal{S}^{p-1}} \Psi^2(f,\mathbf{x})d\sigma_{p-1}(\mathbf{x})$. Minimizing (3.13) with respect to h is complicated due to $c_{h,p}(K_s)$. We therefore suggest replacing $(c_{h,p}(K_s))^{-1}$ with the asymptotically equivalent expression (see (3.6))

$$h^{p-1}\lambda_p(K_s) := h^{p-1} 2^{(p-3)/2}\omega_{p-1}\int_0^\infty K_s(v)v^{(p-3)/2}dv. \quad (3.14)$$

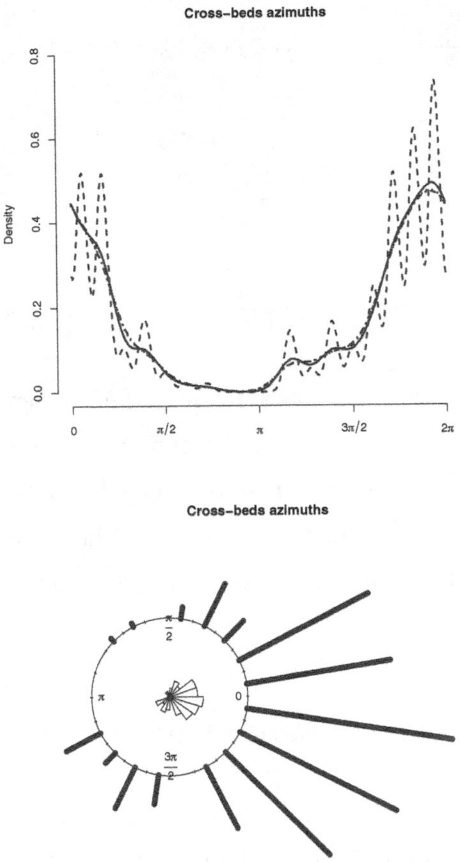

Figure 3.1: Circular kernel density estimators (top) for 580 azimuths of cross-beds in the Kamthi river, an asymmetric dataset as illustrated via a rose diagram (bottom). The bandwidth selection methods are mixtures of von Mises distributions à la Oliveira et al. (2012) (solid line), likelihood cross-validation (dashed line) and Taylor's rule of thumb based on a single von Mises distribution (dotted line). We thank the authors from Oliveira et al. (2012) to provide us with these pictures.

This substitution is possible since the AMISE is the asymptotic part of the MISE. Minimizing (3.13) with respect to h is now straightforward and yields

$$h_{\text{AMISE}} = \left(\frac{(p-1)d_p(K_s)}{4(b_p(K_s))^2 \lambda_p(K_s) R(\boldsymbol{\Psi}, f) n} \right)^{\frac{1}{3+p}} . \tag{3.15}$$

This expression cannot be used in practice, since it depends on the unknown density f through the factor $R(\Psi, f)$. The following sections present ways to circumvent this issue.

3.3.2 Rule of thumb based on the FvML distribution

As explained in Chapter 2, the Fisher–von Mises–Langevin distributions are spherical analogues of the normal distribution. It is hence not surprising that Silverman (1986)'s rule of thumb, based on a normal kernel and normal reference density (see Section 3.1), can be adapted to the spherical setting through an FvML kernel and an FvML reference density f.

García-Portugués (2013) showed that the curvature term $R(\Psi, f)$ for f the FvML density with location $\boldsymbol{\mu} \in \mathcal{S}^{p-1}$ and concentration $\kappa \geq 0$ is given by

$$\frac{\kappa^{p/2}}{2^{p+1}\pi^{p/2}(\mathcal{I}_{(p-2)/2}(\kappa))^2(p-1)}\left(2(p-1)\mathcal{I}_{p/2}(2\kappa)+(p+1)\kappa\mathcal{I}_{(p+2)/2}(2\kappa)\right). \tag{3.16}$$

Substituting a suitable estimator $\hat{\kappa}$ for the concentration parameter then leads to the following automatic bandwidth expression for general kernel K_s:

$$h_{\mathrm{AUTO}} = \left(\frac{(p-1)^2 d_p(K_s) 2^{p+1}\pi^{p/2}(\mathcal{I}_{(p-2)/2}(\hat{\kappa}))^2}{\hat{\kappa}^{p/2}4(b_p(K_s))^2\lambda_p(K_s)\left(2(p-1)\mathcal{I}_{p/2}(2\hat{\kappa})+(p+1)\hat{\kappa}\mathcal{I}_{(p+2)/2}(2\hat{\kappa})\right)n}\right)^{\frac{1}{3+p}}$$

where $\lambda_p(K_s)$ is still defined as in (3.14). With the popular FvML kernel $K_s(v) = e^{-v}$, the different factors involved simplify into $b_p(\mathrm{FvML}) = (p-1)/2$, $d_p(\mathrm{FvML}) = 2^{(1-p)/2}$ and $\lambda_p(\mathrm{FvML}) = (2\pi)^{(p-1)/2}$, and one can readily simplify h_{AUTO}. For the sake of comparison with what follows, we explicitly give this selector in the circular setting:

$$h_{\mathrm{AUTO;FvML}} = \left(\frac{4\pi^{1/2}(\mathcal{I}_0(\hat{\kappa}))^2}{\hat{\kappa}\left(2\mathcal{I}_1(2\hat{\kappa})+3\hat{\kappa}\mathcal{I}_2(2\hat{\kappa})\right)n}\right)^{1/5}.$$

Taylor (2008) was the first to propose an automatic bandwidth selector for the circular case. His approach slightly differs from Silverman (1986) in that he calculated bias and variance expressions, and hence the AMISE, by replacing from the beginning the unknown f with the von Mises density, and not only at the level of $R(\Psi, f)$. This leads to a change in the AMISE expression, and consequently to the following automatic bandwidth selector:

$$h_{\mathrm{TAY}} = \left(\frac{4\pi^{1/2}(\mathcal{I}_0(\hat{\kappa}))^2}{\hat{\kappa}^2 3\mathcal{I}_2(2\hat{\kappa})n}\right)^{1/5}.$$

The difference between $h_{\mathrm{AUTO;FvML}}$ and h_{TAY} is the term $2I_1(2\hat{\kappa})$ in the denominator of $h_{\mathrm{AUTO;FvML}}$. The bandwidth h_{TAY} is more sensible to departures from the von Mises assumption for f than $h_{\mathrm{AUTO;FvML}}$.

3.3.3 A gain in generality: AMISE via mixtures of FvML densities

Despite being less von Mises dependent than h_{TAY}, the automatic bandwidth selector $h_{\mathrm{AUTO;FvML}}$ still suffers from a lack of flexibility due to replacing f with the FvML density when calculating the curvature term (see (3.16)). The FvML density is not versatile enough to capture densities with multimodality and it approximates them by the flat uniform density. A natural way to overcome this limitation is to calculate $R(\boldsymbol{\Psi}, f)$ when f is a more flexible density, see Chapter 2. Oliveira et al. (2012) and García-Portugués (2013) both opted for mixtures of FvML densities, with the number of components M depending on the data under investigation. Mixtures of M FvML densities $f^{\mathrm{FvML}}(\mathbf{x}; \boldsymbol{\mu}_i, \kappa_i)$ with respective location $\boldsymbol{\mu}_i \in \mathcal{S}^{p-1}$ and concentration $\kappa_i \geq 0$, $i = 1, \ldots, n$, have densities of the form

$$f_M^{\mathrm{FvML}}(\mathbf{x}; \boldsymbol{\mu}_1, \ldots, \boldsymbol{\mu}_M, \kappa_1, \ldots, \kappa_M) := \sum_{i=1}^{M} p_i f^{\mathrm{FvML}}(\mathbf{x}; \boldsymbol{\mu}_i, \kappa_i),$$

with

$$\sum_{i=1}^{M} p_i = 1,\ p_i \geq 0\ \forall i = 1, \ldots, n.$$

A drawback of this gain in generality however is that the curvature term $R(\boldsymbol{\Psi}, f_M^{\mathrm{FvML}})$ can no longer be expressed in closed form when $M > 1$, and needs to be computed numerically. This can be done either by numerical or Monte Carlo integration methods. Using the general h_{AMISE} expression in (3.15), bandwidth selectors based on mixtures of FvML distributions are then constructed in two steps:

1. From the observations $\mathbf{X}_1, \ldots, \mathbf{X}_n \in \mathcal{S}^{p-1}$, choose the best-fitting mixture

$$f_{\hat{M}}(\mathbf{x}) := f_{\hat{M}}^{\mathrm{FvML}}(\mathbf{x}; \hat{\boldsymbol{\mu}}_1, \ldots, \hat{\boldsymbol{\mu}}_{\hat{M}}, \hat{\kappa}_1, \ldots, \hat{\kappa}_{\hat{M}})$$

   with $\hat{M}$ the estimated number of components.

2. Compute numerically $R(\boldsymbol{\Psi}, f_{\hat{M}})$ and hence

$$h_{\mathrm{AMI}} = \left(\frac{(p-1)d_p(K_s)}{4(b_p(K_s))^2 \lambda_p(K_s) R(\boldsymbol{\Psi}, f_{\hat{M}}) n} \right)^{\frac{1}{3+p}}$$

   where AMI refers to Asymptotic MIxtures.

Step 1 requires some comments. For a fixed number of components M, the parameters $\boldsymbol{\mu}_1, \ldots, \boldsymbol{\mu}_M$ and $\kappa_1, \ldots, \kappa_M$ can be estimated using an Expectation-Maximization (EM) algorithm described in Banerjee et al. (2005). The most suitable number of components, $\hat{M}$, can then be found by means of some information criterion, e.g., the Akaike Information Criterion or the Bayesian Information Criterion.

This plug-in rule, based on FvML mixtures instead of the single FvML distribution when calculating the curvature term, improves on the rule of thumb selectors of the previous section in terms of generality and robustness towards departures from the FvML assumption. This fact has been corroborated in a thorough simulation study reported in García-Portugués (2013).

In the circular case, Oliveira et al. (2012) proceeded along similar lines with a mixture of von Mises densities. They however used a different AMISE expression, namely that of Di Marzio et al. (2009) based on Fourier expansion of the circular kernel.

3.3.4 Three further proposals

Minimizing the AMISE is, of course, not the only automatic bandwidth selection method. Further proposals have been put forward in the literature, and we shall briefly mention three of them.

The first two proposals stem from Hall et al. (1987) and are based on cross-validation. For a sample of n independent observations $\mathbf{X}_1, \ldots, \mathbf{X}_n \in \mathcal{S}^{p-1}$, the Least Squares Cross-Validation selector h_{LSCV} minimizes the squared error loss if it maximizes

$$\text{CV}_2(h) = \frac{2}{n} \sum_{i=1}^{n} \hat{f}_h^{-i}(\mathbf{X}_i) - \int_{\mathcal{S}^{p-1}} (\hat{f}_h(\mathbf{x}))^2 d\sigma_{p-1}(\mathbf{x})$$

where $\hat{f}_h^{-i}$ stands for the kernel density estimator obtained without the i-th observation. In a similar way, the Likelihood Cross-Validation selector h_{LCV} minimizes the Kullback–Leibler loss if it maximizes

$$\text{CV}_{\text{KL}} = \sum_{i=1}^{n} \log \hat{f}_h^{-i}(\mathbf{X}_i).$$

The third proposal is due to García-Portugués (2013) and is grounded on the idea of minimizing (with respect to h) the MISE instead of the AMISE. This results in a gain in accuracy for small to moderate sample sizes, where the AMISE can differ substantially from the MISE; in turn, it also requires more involved calculations. For a mixture of FvML densities as reference density f and von Mises kernel K_s, García-Portugués et al. (2013*b*) provided a non-trivial closed-form expression for the MISE. The related bandwidth parameter is denoted by h_{EMI} for Exact MIxtures.

3.3.5 Bandwidth selection in the cylindrical setting

Bandwidth selection in cylindrical settings is far more convoluted than on hyperspheres, given that a spherical as well as a linear bandwidth need to be determined. The AMISE expression can be derived from (3.11) and (3.12). Getting closed-form expressions for $(h, g)_{\mathrm{AMISE}}$ is cumbersome, and numerical optimization tools must be used. García-Portugués et al. (2013*b*) therefore considered the special case where $g = \beta h$ for some $\beta > 0$. Under such a proportionality assumption, they derived h_{AMISE} and provided a closed-form expression for β in the circular-linear setting.

3.4 Inferential procedures

Besides density estimation *per se*, kernel density estimation methods pave also the way for various non-parametric inferential procedures. In this section, we shall describe a selection of recent advances in this direction.

3.4.1 Non-parametric goodness-of-fit test for directional data

Consider a sample of n iid observations $\mathbf{X}_1, \ldots, \mathbf{X}_n$ on $\mathcal{S}^{p-1}$, and suppose we are interested in identifying the unknown common density f of the $\mathbf{X}_i$'s. We shall distinguish here two situations: a simple null hypothesis $\mathcal{H}_0 : f = f_0$ for some fixed density f_0 (for examples, see Section 2.3 of Chapter 2) and a composite hypothesis $\mathcal{H}_0 : f \in \mathcal{F} = \{f_{\boldsymbol{\vartheta}}\}$ for $\boldsymbol{\vartheta}$ some vector of parameters (e.g., location and concentration for FvML densities). In the latter scenario, the parameter $\boldsymbol{\vartheta}$ needs to be estimated by some estimator $\hat{\boldsymbol{\vartheta}}$ under the null hypothesis.

In both settings, a non-parametric goodness-of-fit test can be based on some distance between f_0, respectively $f_{\hat{\boldsymbol{\vartheta}}}$, and the kernel density estimate $\hat{f}_h$ from (3.4). Typical choices for distances are the L^p distances, prominently the L^2 distance. For instance, Zhao & Wu (2001) studied the asymptotic behavior of the L^2 distance between f_0 and $\hat{f}_h$. However, it is advisable not to compare $\hat{f}_h$ with f_0, respectively $f_{\hat{\boldsymbol{\vartheta}}}$, directly, because of the inherent bias of $\hat{f}_h$. Depending on the choice of the bandwidth h, this may cause serious power loss when the bias becomes large. Consequently, it is better to define test statistics based on the distance between $\hat{f}_h$ and its estimated value under the null hypothesis, which is a smooth version of f_0, respectively $f_{\hat{\boldsymbol{\vartheta}}}$. This respectively leads to consider distances of the form

$$T_{s,n} = \int_{\mathcal{S}^{p-1}} \left(\hat{f}_h(\mathbf{x}) - K_h(f_0(\mathbf{x}))\right)^2 d\sigma_{p-1}(\mathbf{x})$$

and

$$T_{c,n} = \int_{\mathcal{S}^{p-1}} \left(\hat{f}_h(\mathbf{x}) - K_h(f_{\hat{\boldsymbol{\vartheta}}}(\mathbf{x}))\right)^2 d\sigma_{p-1}(\mathbf{x}),$$

where $K_h(\psi(\mathbf{x})) = c_{h,p}(K_s) \int_{\mathcal{S}^{p-1}} K_s\left(\frac{1-\mathbf{x}'\mathbf{y}}{h^2}\right) \psi(\mathbf{y}) d\sigma_{p-1}(\mathbf{y})$ is the expected value of $\hat{f}_h(\mathbf{x})$ under the density ψ. Tests based on $T_{s,n}$ and $T_{c,n}$ were investigated in Boente et al. (2014). Assuming SKDE1-SKDE3 to hold, and that, for $T_{c,n}$, (i) $\boldsymbol{\vartheta} \mapsto f_{\boldsymbol{\vartheta}}$ is twice continuously differentiable with bounded and uniformly continuous (in $(\boldsymbol{\vartheta}, \mathbf{x})$) partial derivatives, and (ii) $\hat{\boldsymbol{\vartheta}}$ is a root-n consistent estimator for $\boldsymbol{\vartheta}$, they proved the asymptotic normality of $T_{s,n}$ and $T_{c,n}$ at convergence rate $nh^{(p-1)/2}$.

Given this rather slow rate of convergence, one can expect that the normal approximation does not work well for small or moderate sample sizes. Consequently, bootstrap versions of the tests based on $T_{s,n}$ and $T_{c,n}$ should be considered. These can be built in three steps:

1. Generate a random sample $\mathbf{X}_1^*, \ldots, \mathbf{X}_n^*$ of size n from f_0, respectively $f_{\hat{\boldsymbol{\vartheta}}}$, for some root-$n$ consistent estimator $\hat{\boldsymbol{\vartheta}}$.

2. Compute $T_{s,n}^* = \int_{\mathcal{S}^{p-1}} \left(\hat{f}_h^*(\mathbf{x}) - K_h(f_0(\mathbf{x}))\right)^2 d\sigma_{p-1}(\mathbf{x})$ and

$$T_{c,n}^* = \int_{\mathcal{S}^{p-1}} \left(\hat{f}_h^*(\mathbf{x}) - K_h(f_{\hat{\boldsymbol{\vartheta}}^*}(\mathbf{x}))\right)^2 d\sigma_{p-1}(\mathbf{x}),$$

   where the estimates $\hat{f}_h^*(\mathbf{x})$ and $\hat{\boldsymbol{\vartheta}}^*$ are obtained from the sample $\mathbf{X}_1^*, \ldots, \mathbf{X}_n^*$.

3. Repeat Steps 1 and 2 a total of B times and establish the empirical distribution of the sequences $(T_{s,n}^*)_1, \ldots, (T_{s,n}^*)_B$, respectively $(T_{c,n}^*)_1, \ldots, (T_{c,n}^*)_B$.

The rejection rules for $T_{s,n}$ and $T_{c,n}$ are then based on quantiles derived from the empirical distributions of Step 3.

Similar procedures for cylindrical data were proposed in García-Portugués et al. (2015), in conjunction with a Central Limit Theorem for the Integrated Squared Error of (3.10).

3.4.2 Non-parametric independence test for cylindrical data

In the same vein as the goodness-of-fit tests of the previous section, non-parametric tests for spherical-linear independence were studied in García-Portugués et al. (2014) and García-Portugués et al. (2015). Retaining the notation $f(\mathbf{x}, z)$ for the common density of the iid observations $(\mathbf{X}_1, Z_1), \ldots, (\mathbf{X}_n, Z_n)$, the null hypothesis of independence can be written as

$$\mathcal{H}_0 : f(\mathbf{x}, z) = f_{\mathbf{X}}(\mathbf{x}) f_Z(z) \, \forall (\mathbf{x}, z) \in \mathcal{S}^{p-1} \times \mathbb{R}$$

where $f_{\mathbf{X}}(\mathbf{x})$ and $f_Z(z)$ denote the spherical and linear marginal densities, respectively. The alternative hypothesis reads $\mathcal{H}_1 : f(\mathbf{x}, z) \neq f_{\mathbf{X}}(\mathbf{x}) f_Z(z)$ for some $(\mathbf{x}, z) \in \mathcal{S}^{p-1} \times \mathbb{R}$.

A natural non-parametric test for independence is based on the L^2 distance between the cylindrical kernel density estimate $\hat{f}_{h,g}$ of (3.10) and the product $\hat{f}_{\mathbf{X};h}\hat{f}_{Z;g}$ of the marginal kernel density estimates $\hat{f}_{\mathbf{X};h}$ and $\hat{f}_{Z;g}$, respectively defined in (3.4) and (3.1). The corresponding test statistic

$$T_{ind,n} = \int_{\mathcal{S}^{p-1} \times \mathbb{R}} \left(\hat{f}_{h,g}(\mathbf{x}, z) - \hat{f}_{\mathbf{X};h}(\mathbf{x}) \hat{f}_{Z;g}(z) \right)^2 d\sigma_{p-1}(\mathbf{x}) dz$$

is asymptotically normal under the null hypothesis with convergence rate $nh^{(p-1)/2} g^{1/2}$, subject to the usual assumptions plus the additional assumption that h_n^{p-1}/g_n converges to $0 < c < \infty$ as $n \to \infty$. Computing $T_{ind,n}$ is not an easy task, except when using a spherical FvML kernel and a linear normal kernel.

Like all omnibus tests based on smoothing, the test for independence suffers from slow rates of convergence. Instead of a bootstrap version, the null hypothesis of independence suggests the application of the permutation principle. The latter advocates to first recalculate the test statistic for B samples $\{(\mathbf{X}_i, Z_{per_b(i)}), i = 1, \ldots, n\}$, where per_b is a permutation of $\{1, \ldots, n\}$ for $b = 1, \ldots, B$, and then compare the original $T_{ind,n}$ to the resulting empirical quantile. We refer the reader to Hallin & Ley (2012a) for a general review of permutation tests.

3.4.3 An overview of non-parametric regression

Regression models with a spherical response and/or spherical predictor have been studied a lot over the past decades. Notable recent contributions in this area are the seminal papers by Presnell et al. (1998), Downs & Mardia (2002) and Kato et al. (2008). Moreover, each new toroidal or cylindrical distribution (see Section 2.4) gives rise to a regression model. A detailed account of spherical regression models could be the topic of an entire book. Here we shall therefore only briefly mention relevant advances in non-parametric kernel-based regression involving a spherical response or predictor. In mathematical terms, we are interested in regression models of the form $\mathbf{W}_1 = m(\mathbf{W}_2) + \sigma(\mathbf{W}_2)\boldsymbol{\epsilon}$ where $\mathbf{W}_1$ and $\mathbf{W}_2$ are linear or spherical random variables/vectors, the error $\boldsymbol{\epsilon}$ is of the same type as $\mathbf{W}_1$ and independent of $\mathbf{W}_2$, $\sigma^2(\cdot)$ is the conditional variance of $\mathbf{W}_1$ and the unknown function m contains the dependence structure between $\mathbf{W}_1$ and $\mathbf{W}_2$. The following settings have been investigated over the past decade:

- *linear response and circular or toroidal predictor*: Di Marzio et al. (2009) considered W_1 a real-valued scalar response and $\mathbf{W}_2$ a vector of p angles, and a real-valued error ϵ with mean zero and unit variance. They proposed local linear kernel estimates for the unknown function $m(\cdot)$ by minimizing a weighted sum of squares, the weights being based on a product of circular kernels. Deschepper et al. (2008) investigated the setting where W_2 is circular, $\sigma(\cdot) = 1$ and the errors have mean zero and constant variance. They proposed a graphical tool to assess the fit of a certain choice of m, accompanied by a formal lack-of-fit test.

- *linear response and spherical predictor*: this setup is similar to the previous one, with $\mathbf{W}_2$ now taking values on $\mathcal{S}^{p-1}$. Di Marzio et al. (2014) modified the approach of Di Marzio et al. (2009) for linear-toroidal regression to the linear-spherical setting by relying on the tangent-normal decomposition (2.21) for $\mathbf{W}_2$ and using spherical kernels. A different approach was taken by García-Portugués et al. (2017) who do not rely on a tangent-normal decomposition but directly extend $m(\mathbf{W}_2)$ into a Taylor series expansion and then minimize a weighted sum of squares. Whenever W_2 is circular, their approach coincides with that of Di Marzio et al. (2009). The primary aim of García-Portugués et al. (2017), however, does not lie in estimating m but rather on goodness-of-fit tests for certain parametric choices of m.

- *circular response and linear or circular predictor*: Di Marzio et al. (2013) considered W_1 to lie on the unit circle $\mathcal{S}^1$ and W_2 to be either defined on $\mathcal{S}^1$ or on the interval $[0, 1]$. The circular nature of the response is taken account of by estimating $m(\cdot)$ via the arctangent of the ratio of sample statistics based on weighted sines and cosines, the weights depending on circular, respectively linear, kernels. The same problem is also addressed in Di Marzio et al. (2016) by means of quantile regression.

- *spherical response and spherical predictor*: Di Marzio et al. (2014) considered $\mathbf{W}_1$ and $\mathbf{W}_2$ to take values on unit hyperspheres with potentially different dimensions. The regression function m is estimated as the solution to a weighted least squares problem.

3.5 Further reading

In this chapter we have discussed advances in directional kernel density estimation, amended by some modern inferential procedures. Particular focus was on the crucial issue of bandwidth selection. We now complement the preceding pages by providing the reader with references to further interesting work published in recent years.

Estimation of density-related quantities

In the preceding sections we have focussed on the estimation of the probability density function itself. Density-related quantities such as distribution functions or derivatives of densities have as well been studied in recent years. Klemelä (2000) investigated estimation of the Laplacian of densities on $\mathcal{S}^{p-1}$ as well as other types of derivatives. Circular distribution functions were estimated by kernel methods in Di Marzio et al. (2012). The same paper also proposed non-parametric estimates for the related circular quantiles. Conditional circular quantiles, when the conditioning variable can be linear or circular, were studied in Di Marzio et al. (2016).

Cylindrical kernel density estimation exploiting copula structures

Kernel density estimation for cylindrical data was considered in Section 3.2.2. For the particular case of circular-linear data, the same problem has been tackled from a different angle, namely by using the copula-like structure described in Section 2.4.5. For iid observations $(\Theta_1, Z_1), \ldots, (\Theta_n, Z_n)$, García-Portugués et al. (2013*a*) proceeded in three steps: (i) estimate, via kernel density estimation methods, the circular and linear marginal densities f_1 and f_2 from (2.30), and the corresponding distribution functions F_1 and F_2; (ii) compute an artificial sample $(\hat{F}_1(\Theta_i), \hat{F}_2(Z_i)), i = 1, \ldots, n$, and estimate the copula structure c; (iii) the circular-linear density estimate is then calculated as $\hat{c}\left(\hat{F}_1(\theta), \hat{F}_2(z)\right) \hat{f}_1(\theta)\hat{f}_2(z)$. If the copula c is of the simpler form (2.31) with binding function c_b a circular density, then circular kernel density estimation can be employed in (ii). Otherwise the copula can be estimated non-parametrically by adapting to the circular setting the kernel density methods of Gijbels & Mielniczuk (1990). At the same level of generality, Carnicero et al. (2013) used the versatile non-parametric Bernstein copulas in order to estimate the copula structure.

Computational and graphical methods

This chapter is a collection of computational and graphical methods that have appeared in the literature in recent years. Its structure differs from that used in the other chapters, in that each section contains its own introduction with links to procedures from $\mathbb{R}^p$, and then describes the corresponding advances in directional statistics.

4.1 Ordering data on the sphere: quantiles and depth functions

4.1.1 Ordering on $\mathbb{R}$ and $\mathbb{R}^p$, and organization of the remainder of the section

Ordering data on the real line is simple and the concept of quantiles is extremely popular. Basic features of a given data set such as location, dispersion, skewness and kurtosis are typically described by the median, interquartile range, Bowley coefficient of skewness and Moor coefficient of kurtosis. The QQ-plot is a simple and widely used graphical tool to assess the adequacy of a parametric model for the data at hand. Numerous robust inferential procedures rely on quantiles, such as quantile regression (Koenker 2005), quantile goodness-of-fit tests (LaRiccia 1991) or parameter estimation by quantile matching (Dominicy & Veredas 2013). Defining quantiles in $\mathbb{R}^p$ is a much more delicate issue, as there is no canonical means of ordering multivariate observations.

Providing such a multivariate ordering is the goal of *depth functions*. These functions provide a center-outward ordering for any data set by assigning to every point $\mathbf{z} \in \mathbb{R}^p$ a value measuring its centrality within the data cloud. Numerous concepts of depth exist, reflecting the aforementioned lack of a natural ordering in $\mathbb{R}^p$. Two well-known occurrences are Tukey's half-space depth, defined for each point $\mathbf{z}$ as the minimum probability mass carried by closed half-spaces containing

$\mathbf{z}$, and the simplicial depth of Liu (1990), defined as the proportion of data-based simplices containing $\mathbf{z}$. Data-based simplices are simplices whose $p+1$ vertices are all drawn from the data set. A systematic treatment of depth functions is provided by Zuo & Serfling (2000) who set up a list of four desirable properties a statistical depth function should satisfy: (1) affine-invariance, i.e., the depth should not depend on the underlying coordinate system, (2) maximality at the center, for symmetric cases, (3) monotonicity relative to the deepest point, i.e., the depth of $\mathbf{z}$ should be decreasing as $\mathbf{z}$ moves away from the unique deepest point along any ray from that point, and (4) vanishing at infinity, i.e., when $||\mathbf{z}|| \to \infty$.

The remainder of this section is organized as follows. After a brief description in Section 4.1.2 of depth functions introduced around the beginning of the 1990s, we focus, in Sections 4.1.3 and 4.1.4, on the very recent quantile and depth concepts introduced in Ley et al. (2014). Related inferential procedures, descriptive statistics and exploratory data analysis tools are discussed in Section 4.1.5.

4.1.2 Classical depth functions on the sphere

The first notions of depth functions on the sphere can be traced back to Small (1987) and Liu & Singh (1992). Writing $P_f(A)$ the probability measure of $A \subseteq \mathcal{S}^{p-1}$ for a probability law with density f on $\mathcal{S}^{p-1}$, these papers introduced the following depth functions:

- Angular Tukey's depth (Small 1987, Liu & Singh 1992): for a spherical density f, the angular Tukey's depth of a point $\mathbf{x} \in \mathcal{S}^{p-1}$ is defined as

$$ATD_f(\mathbf{x}) = \inf_{\mathcal{H}:\mathbf{x}\in\mathcal{H}} P_f(\mathcal{H}),$$

  where $\mathcal{H}$ is a closed hemisphere of $\mathcal{S}^{p-1}$ containing $\mathbf{x}$ either on its boundary or in its interior.

- Angular simplicial depth (Liu & Singh 1992): for a spherical density f, the angular simplicial depth of a point $\mathbf{x} \in \mathcal{S}^{p-1}$ is defined as

$$ASD_f(\mathbf{x}) = P_f(\mathbf{x} \in \mathrm{Simp}(\mathbf{X}_1, \mathbf{X}_2, \ldots, \mathbf{X}_p)),$$

  where $\mathrm{Simp}(\mathbf{X}_1, \mathbf{X}_2, \ldots, \mathbf{X}_p)$ stands for the simplex on $\mathcal{S}^{p-1}$ with vertices $\mathbf{X}_1, \ldots, \mathbf{X}_p$ that are iid random vectors with density f.

- Arc distance depth (Liu & Singh 1992): for a spherical density f, the arc distance depth of a point $\mathbf{x} \in \mathcal{S}^{p-1}$ is defined as

$$ADD_f(\mathbf{x}) = \pi - \int_{\mathcal{S}^{p-1}} \ell(\mathbf{x}, \mathbf{y}) f(\mathbf{y}) d\sigma_{p-1}(\mathbf{y}),$$

  where $\ell(\mathbf{x}, \mathbf{y})$ stands for the length of the short arc joining $\mathbf{x}$ and $\mathbf{y}$ on the great circle containing $\mathbf{x}$ and $\mathbf{y}$.

The empirical versions of ATD and ADD are defined by replacing P_f with the empirical probability measure of the random observations $\mathbf{X}_1, \ldots, \mathbf{X}_n \in \mathcal{S}^{p-1}$, while the empirical angular simplicial depth is calculated as

$$\frac{1}{\binom{n}{p}} \sum_{1 \leq i_1 < i_2 < \ldots < i_p \leq n} \mathbb{I}[\mathbf{x} \in \text{Simp}(\mathbf{X}_{i_1}, \mathbf{X}_{i_2}, \ldots, \mathbf{X}_{i_p})],$$

where the sum runs over all p-tuples $1 \leq i_1 < i_2 < \ldots < i_p \leq n$.

These classical depth functions provide center-outward ordering of the data on spheres and can be used to define various spherical medians. However, asymptotic normality or asymptotic representation results, and hence inferential procedures based on these data depths, are extremely difficult to obtain. Moreover, the angular Tukey's depth suffers from the unpleasant feature that all points of an entire hemisphere (the one with the lowest probability measure) are given the same depth.

4.1.3 Projected quantiles and related asymptotic results

Ley et al. (2014) introduced a novel concept of spherical quantiles which we call *projected quantiles* for reasons that will shortly become obvious.

Consider the class $\mathcal{F}$ of probability distributions on $\mathcal{S}^{p-1}$ that have a bounded density and a unique median direction. This rules out antipodally symmetric distributions, for which quantiles would anyhow make little sense. The choice of median is arbitrary, hence we simply denote it by $\boldsymbol{\mu}_m$ without commenting on its concrete definition. Note that, for rotationally symmetric distributions (Section 2.3.2), various medians such as the Fisher median (Section 5.3.2) or the deepest points of the angular Tukey's or simplicial depths all coincide with the unique mode. Let $\mathbf{X}$ be a random vector on $\mathcal{S}^{p-1}$ following a probability law in $\mathcal{F}$ and consider the quantile check function $\rho_\tau(z) := z(\tau - \mathbb{I}[z \leq 0])$, $z \in \mathbb{R}$, $\tau \in [0, 1]$. Define

$$c_\tau := \arg \min_{c \in [-1,1]} \mathrm{E}\left[\rho_\tau(\mathbf{X}'\boldsymbol{\mu}_m - c)\right] \tag{4.1}$$

as the univariate quantile of order τ obtained after projecting the vector $\mathbf{X}$ onto the median $\boldsymbol{\mu}_m$. The quantity c_τ is termed the τth *projection quantile* and the related τth quantile is thus $c_\tau\boldsymbol{\mu}_m$, hence the name projected quantile. The value $\tau = 1$, for which $c_1 = 1$, is attained by $\boldsymbol{\mu}_m$, while $\tau = 0$, for which $c_0 = -1$, is attained by $-\boldsymbol{\mu}_m$ provided the neighborhood of $-\boldsymbol{\mu}_m$ has a positive probability mass. Otherwise, an entire cap centered at $-\boldsymbol{\mu}_m$ will be assigned the value $\tau = 0$. Furthermore, each c_τ leads to the subsets

$$\mathcal{C}_\tau^+ := \{\mathbf{x} \in \mathcal{S}^{p-1} \,|\, \mathbf{x}'\boldsymbol{\mu}_m \geq c_\tau\} \quad \text{and} \quad \mathcal{C}_\tau^- := \{\mathbf{x} \in \mathcal{S}^{p-1} \,|\, \mathbf{x}'\boldsymbol{\mu}_m < c_\tau\}$$

which are upper and lower quantile caps for $\mathbf{X}$. These quantile caps provide a simple graphical means of assessing the spread of the data distribution around the median direction. We illustrate this statement in Figure 4.1, which portrays upper quantile caps for a concentrated ($\kappa = 10$) and a less concentrated ($\kappa = 2$) FvML distribution.

The empirical version of the projected quantiles for observations $\mathbf{X}_1, \ldots, \mathbf{X}_n$ is built in three steps:

1. compute a root-n consistent empirical median $\hat{\boldsymbol{\mu}}_m$;

2. compute the projections $\mathbf{X}_1'\hat{\boldsymbol{\mu}}_m, \ldots, \mathbf{X}_n'\hat{\boldsymbol{\mu}}_m$;

3. calculate
$$\hat{c}_\tau := \arg\min_{c\in[-1,1]} n^{-1}\sum_{i=1}^{n} \rho_\tau(\mathbf{X}_i'\hat{\boldsymbol{\mu}}_m - c).$$

The resulting projected quantile of order τ corresponds to $\hat{c}_\tau\hat{\boldsymbol{\mu}}_m$. It is rotation-equivariant by construction. Letting f_{proj} denote the common density of the $\mathbf{X}_i'\boldsymbol{\mu}_m$'s, Ley et al. (2014) showed that there exists a p-vector $\boldsymbol{\Gamma}_{\boldsymbol{\mu}_m, c_\tau}$ such that[1]

$$n^{1/2}(\hat{c}_\tau - c_\tau) = \frac{n^{-1/2}}{\Gamma_{c_\tau}}\sum_{i=1}^{n}(\tau - \mathbb{I}[\mathbf{X}_i'\boldsymbol{\mu}_m \leq c_\tau]) - \frac{\boldsymbol{\Gamma}'_{\boldsymbol{\mu}_m,c_\tau}}{\Gamma_{c_\tau}} n^{1/2}(\hat{\boldsymbol{\mu}}_m - \boldsymbol{\mu}_m) + o_{\mathrm{P}}(1) \tag{4.2}$$

as $n \to \infty$, where $\Gamma_{c_\tau} := f_{\text{proj}}(c_\tau)$. This asymptotic result is a Bahadur-type representation of $\hat{c}_\tau$, and simplifies considerably when the observations are drawn

[1]In fact, the derivation of this asymptotic result requires the use of a locally and asymptotically discrete version of the median $\hat{\boldsymbol{\mu}}_m$; see Section 5.3.3 for details. While important for theoretical developments, this requirement has no impact at finite sample size n.

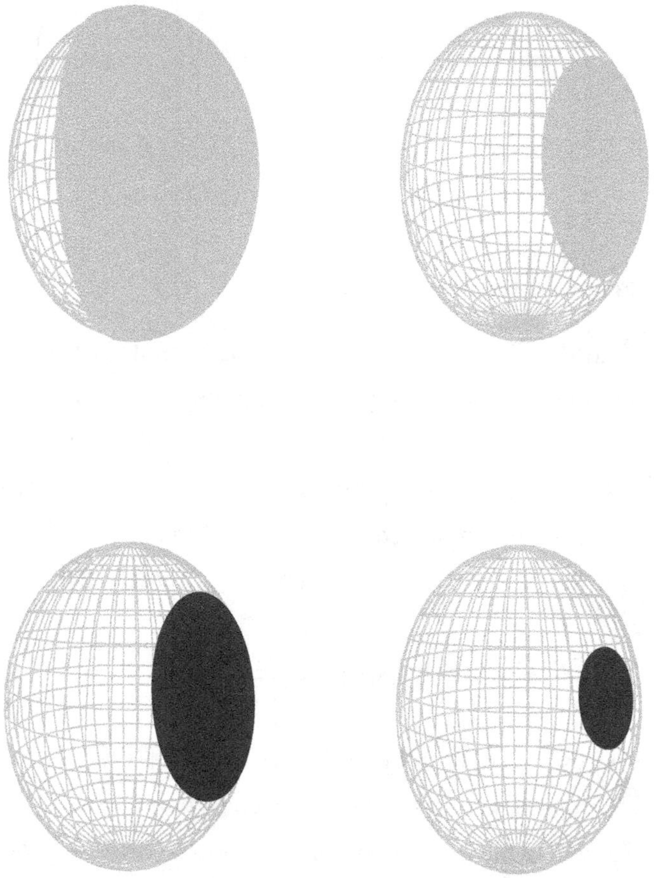

Figure 4.1: Upper quantile caps of order $\tau = 0.05$ and 0.5 for FvML distributions centered at $\boldsymbol{\mu} = (1,0,0)'$ and with respective concentrations $\kappa = 2$ (in grey) and $\kappa = 10$ (in black).

from a rotationally symmetric distribution within $\mathcal{F}$:

$$n^{1/2}(\hat{c}_\tau - c_\tau) = \frac{n^{-1/2}}{\Gamma_{c_\tau}} \sum_{i=1}^{n} (\tau - \mathbb{I}[\mathbf{X}_i'\boldsymbol{\mu}_m \leq c_\tau]) + o_{\mathrm{P}}(1) \qquad (4.3)$$

as $n \to \infty$. This result implies that $n^{1/2}(\hat{c}_\tau - c_\tau)$ is asymptotically normal with mean zero and variance $(1-\tau)\tau/f^2_{\mathrm{proj}}(c_\tau)$. A similar asymptotic normality result can be established using the general representation (4.2) provided one can prove the joint asymptotic normality of

$$\left(n^{-1/2}\sum_{i=1}^{n}(\tau - \mathbb{I}[\mathbf{X}_i'\boldsymbol{\mu}_m - c_\tau \leq 0]), n^{1/2}(\hat{\boldsymbol{\mu}}_m - \boldsymbol{\mu}_m)'\right)'. \tag{4.4}$$

The very simple Bahadur-type representation (4.3) suggests that the projected quantiles are tailor-made for rotationally symmetricdistributions.

4.1.4 The angular Mahalanobis depth

Consider a spherical distribution with density f. Ley et al. (2014) translated the projected quantiles into a depth setting by calculating, for every $\mathbf{x} \in \mathcal{S}^{p-1}$, the quantile value $D_f(\mathbf{x}) := \arg\min_{\tau\in[0,1]}\{c_\tau \geq \mathbf{x}'\boldsymbol{\mu}_m\}$. This value measures the centrality of $\mathbf{x}$ with respect to f and provides a center-outward ordering from the center $\boldsymbol{\mu}_m$. The resulting depth, called the angular Mahalanobis depth because of its similarities with the Mahalanobis depth on $\mathbb{R}^p$, is readily defined in terms of D_f:

$$AMHD_f(\mathbf{x}) := \frac{1}{1+\frac{1}{D_f(\mathbf{x})}}.$$

This rewriting ensures the depth to be of the form $1/(1+O_f(\mathbf{x}))$ for some measure $O_f(\mathbf{x})$ of outlyingness with respect to the deepest point, exactly like the Mahalanobis depth in $\mathbb{R}^p$ where the measure of outlyingness is the Mahalanobis distance from the deepest point. The depth $AMHD_f$ satisfies the four requirements identified in Section 4.1.2. Its contours coincide with the boundary separating upper and lower quantile caps. Finally, the empirical angular Mahalanobis depth is readily obtained by replacing c_τ by $\hat{c}_\tau$ in D_f.

4.1.5 Statistical procedures based on projected quantiles and the angular Mahalanobis depth

We shall now describe various statistical procedures based on the projected quantiles and the angular Mahalanobis depth. Clearly, the quantiles can serve as robust measures of concentration and hence as information-rich complements to the mean resultant length. QQ-plots for spherical distributions are easily constructed and generalize the colatitude plots of Lewis & Fisher (1982) designed to explore the fit of FvML distributions. Four QQ-plots are depicted in Figure 4.2, with target distributions the FvML and the Purkayastha distribution, defined in Section 2.3.2. The depth-based analogue of QQ-plots are DD-plots, which compare theoretical depth

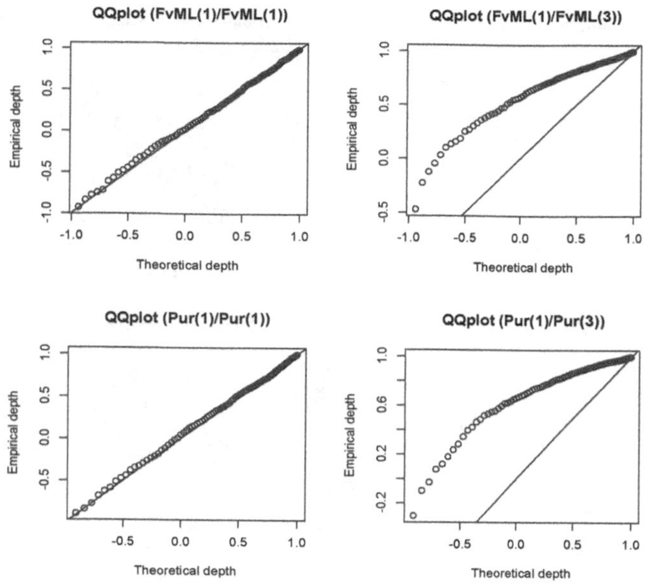

Figure 4.2: QQ-plots (theoretical quantiles versus sample quantiles) using theoretical quantiles from an FvML distribution with concentration 1, denoted FvML(1), in the upper plots and theoretical quantiles from a Purkayastha distribution with concentration 1, denoted Pur(1), in the lower plots. In each case, we generated a sample of 1000 observations from various distributions: for the upper left QQ-plot from an FvML(1) distribution, for the upper right QQ-plot from an FvML(3) distribution, for the lower left QQ-plot from a Pur(1) distribution, and for the lower right QQ-plot from a Pur(3) distribution.

values with their empirical counterparts. The angular Mahalanobis depth allows us to draw such DD-plots, see Figure 4.3. Similar plots can also, of course, be built via the angular Tukey's, simplicial and distance depth functions.

Both the QQ-plots and DD-plots are exploratory tools to assess whether a parametric distribution fits the observations under investigation. A more formal goodness-of-fit test for the problem $\mathcal{H}_0 : f = f_0$ versus $\mathcal{H}_1 : f \neq f_0$ for a rotationally symmetric $f_0 \in \mathcal{F}$ can be constructed thanks to the asymptotic normality of expression (4.3). Consider the statistic $\mathbf{T}_{\boldsymbol{\tau}}^{(n)} := n^{1/2}((\hat{c}_{\tau_1} - c^0_{\tau_1}), \ldots, (\hat{c}_{\tau_r} - c^0_{\tau_r}))'$, where $\boldsymbol{\tau} := (\tau_1, \ldots, \tau_r) \in (0,1)^r$ for $r \in \mathbb{N}$ while the $c^0_{\tau_i}$'s and $\hat{c}_{\tau_i}$'s denote the projection quantiles under f_0 and their empirical counterparts, respectively. Under

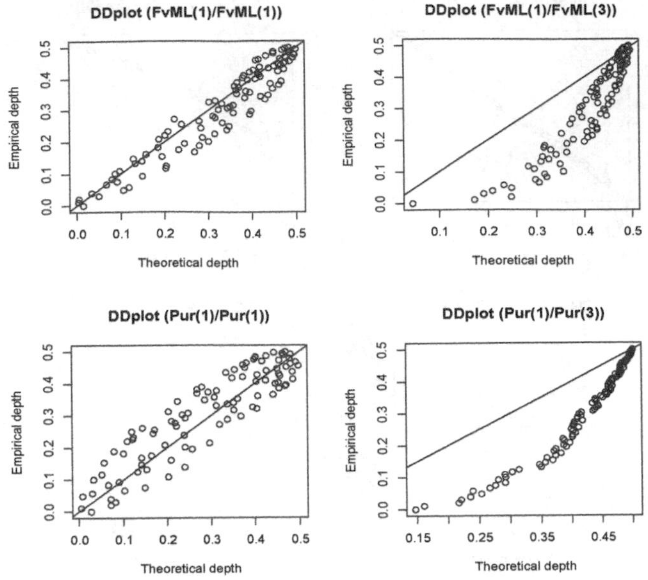

Figure 4.3: DD-plots (theoretical depths versus empirical depths) using theoretical depths from an FvML distribution with concentration 1, denoted FvML(1), in the upper plots and theoretical depths from a Purkayastha distribution with concentration 1, denoted Pur(1), in the lower plots. In each case, we generated a sample of 100 observations from various distributions: for the upper left DD-plot from an FvML(1) distribution, for the upper right DD-plot from an FvML(3) distribution, for the lower left DD-plot from a Pur(1) distribution, and for the lower right DD-plot from a Pur(3) distribution.

$\mathcal{H}_0$, $\mathbf{T}_{\tau}^{(n)}$ is asymptotically normal with mean zero and $r \times r$ covariance matrix

$$\mathbf{\Sigma} = \left(\frac{\min(\tau_i, \tau_j) - \tau_i \tau_j}{f_{0;\mathrm{proj}}(c^0_{\tau_i}) f_{0;\mathrm{proj}}(c^0_{\tau_j})} \right)_{1 \leq i,j \leq r}.$$

The null hypothesis is rejected at asymptotic level α whenever $(\mathbf{T}_{\tau}^{(n)})'\mathbf{\Sigma}^{-1}\mathbf{T}_{\tau}^{(n)}$ exceeds $\chi^2_{r;1-\alpha}$.

4.2 Statistical inference under order restrictions on the circle

4.2.1 Isotonic regression estimation and organization of the remainder of the section

Isotonic regression is a mathematical technique designed to solve the following problem. Given a real-valued sequence $a_1, \ldots, a_q$, find a monotone sequence $\mathbf{a}^m := (a_1^m, \ldots, a_q^m)'$ that best summarizes the information contained in the original sequence. In other words, we want to find

$$\mathbf{a}^m = \text{argmin}_{\mathbf{y} \in \mathbb{R}^q} \sum_{i=1}^{q} (y_i - a_i)^2 \quad \text{under the constraint } y_1 \leq y_2 \leq \cdots \leq y_q. \tag{4.5}$$

The L^2 distance is often used because of the direct link with the sum of squared errors. There are instances where the L^2 distance is replaced with other L^p distances. It can also occur that weights $w_1, \ldots, w_q$ are associated to the summands. Isotonic regression can thus be interpreted as linear regression under order constraints.

As the formulation (4.5) reveals, such problems typically fall under the umbrella of numerical analysis and require operational research tools to be solved. Suppose we have observations $\mathbf{Z}_1, \ldots, \mathbf{Z}_n$ on $\mathbb{R}^p$ and we want to estimate their common location parameter $\boldsymbol{\mu} = (\mu_1, \ldots, \mu_p)' \in \mathbb{R}^p$ under the constraint that $\boldsymbol{\mu} \in \mathcal{L} \subseteq \mathbb{R}^p$ with $\mathcal{L} := \{\mathbf{v} = (v_1, \ldots, v_p)' \in \mathbb{R}^p : v_1 \leq v_2 \leq \cdots \leq v_p\}$. Denoting the unrestricted estimator of $\boldsymbol{\mu}$ by $\hat{\boldsymbol{\mu}}$, the constrained estimator $\hat{\boldsymbol{\mu}}_{\mathcal{L}}$ is found by determining the point in $\mathcal{L}$ that is closest to $\hat{\boldsymbol{\mu}}$ with respect to a given metric. This estimator is called the isotonic regression estimator as it is the solution to (4.5) with $a_i = \hat{\mu}_i, i = 1, \ldots, p$. More complicated constraints can of course also be considered.

Problem (4.5) admits a unique solution that can be found by applying the *pool adjacent violators algorithm (PAVA)*. Order-restricted inferential procedures were first studied in the 1950s, the PAVA algorithm being a major contribution to the field. It can be informally summarized as follows:

- Set $y_i = a_i$ for all $i = 1, \ldots, q$.
- Start from y_1 and move in the sequence of y_i's until the first violation $y_j > y_{j+1}$.

- (Pooling adjacent violators) Replace the pair (y_j, y_{j+1}) with their average, and check if this does not violate monotonicity with respect to y_{j-1}. If not, continue the iterative scheme; if it does, back-average until monotonicity is reached, and then continue the iterative scheme.

- Stop the procedure when the value of y_q is found.

Besides its simplicity, PAVA is popular because of its linear algorithmic complexity. Extensions of PAVA to more complex situations, a description of its computational aspects, and details on the R package for isotonic regression are provided in the recent paper by de Leeuw et al. (2009). Classical reference books dealing with order restricted inference are Barlow et al. (1972) and Robertson et al. (1988).

The remainder of this section is organized as follows. We first explain the geometry underpinning circular order restrictions in Section 4.2.2. Subsequently, in Section 4.2.3, we explain in detail how to estimate parameters subject to order restrictions on the circle.

4.2.2 Order restrictions on the circle

Consider a set of p-dimensional angular observations $\mathbf{\Theta}_1, \ldots, \mathbf{\Theta}_n \in [0, 2\pi)^p$ whose location parameter $\boldsymbol{\mu} = (\mu_1, \ldots, \mu_p)' \in [0, 2\pi)^p$ we wish to estimate under order restrictions. Here we opt for the interval $[0, 2\pi)$ instead of the usual $[-\pi, \pi)$ because it clarifies the statements. Ordering p angles $\mu_1, \ldots, \mu_p$ on the unit circle obviously differs from ordering p real numbers $a_1, \ldots, a_p$: while $a_1 \leq \cdots \leq a_p$ is unambiguous, the same cannot be said of $\mu_1 \leq \cdots \leq \mu_p$ because the actual values of the angles depend on the choice of the origin and the clockwise or anti-clockwise direction used to measure them. An unambiguous *circular order* means that the points $(\cos(\mu_i), \sin(\mu_i))'$, $i = 1, \ldots, p$, follow each other anti-clockwise, in accordance with the standards defined in the Introduction. The circular order therefore is denoted by $\mu_1 \preceq \cdots \preceq \mu_p \preceq \mu_1$ and corresponds to situations like those in Figure 4.4.

Writing $\boldsymbol{\mu} = (\mu_1, \ldots, \mu_p)'$, this order restriction is equivalent to $\boldsymbol{\mu} \in \mathcal{C} := \{\boldsymbol{\phi} \in [0, 2\pi)^p : \phi_1 \preceq \cdots \preceq \phi_p \preceq \phi_1\}$. In terms of classical inequalities, the set $\mathcal{C}$ corresponds to $\cup_{i=1}^p \mathcal{C}^i$ with

$$\mathcal{C}^i = \{\boldsymbol{\phi} \in [0, 2\pi)^p : 0 \leq \phi_i \leq \phi_{i+1} \leq \cdots \leq \phi_{i-1} \leq 2\pi\}, i = 1, \ldots, p.$$

This definition of $\mathcal{C}^i$ is a slight abuse of notation when $i = p$ or $i = 1$; we tacitly put $\phi_{p+1} = \phi_1$ and $\phi_0 = \phi_p$.

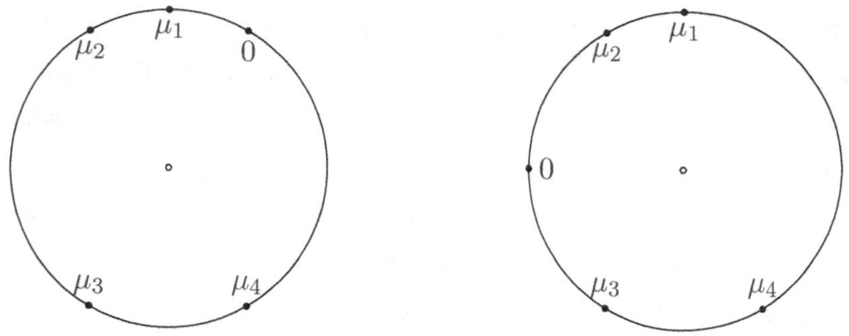

Figure 4.4: Two situations corresponding to the same circular order $\mu_1 \preceq \mu_2 \preceq \mu_3 \preceq \mu_4 \preceq \mu_1$. If one used a non-circular ordering, then the distinct emplacements of the zero angle would lead to two distinct orders.

4.2.3 Circular isotonic regression

Order constrained inference with angular parameters was first considered in Rueda et al. (2009) in relation with phase angles of cell-cycle genes. Recall our goal to estimate angular parameters $\mu_1, \ldots, \mu_p$ under the constraint $\mu_1 \preceq \cdots \preceq \mu_p \preceq \mu_1$, and that we have some unrestricted estimator $\hat{\boldsymbol{\mu}} = (\hat{\mu}_1, \ldots, \hat{\mu}_p)'$ (typically the vector of sample circular mean directions). As in the linear case, the goal is to find $\hat{\boldsymbol{\mu}}^c = (\hat{\mu}_1^c, \ldots, \hat{\mu}_p^c)' \in \mathcal{C}$ that is closest to $\hat{\boldsymbol{\mu}}$ in terms of some distance. In the linear setting, the distance is motivated as sum of squared errors; mimicking this choice in the circular setting, it seems natural to use the sum of circular errors (SCE) given by

$$\mathrm{SCE}(\hat{\boldsymbol{\mu}}, \hat{\boldsymbol{\mu}}^c) = \sum_{i=1}^{p} r_i(1 - \cos(\hat{\mu}_i - \hat{\mu}_i^c))$$

where the r_i's are mean resultant lengths. This circular distance is not new to the attentive reader: it is of the same nature as the angular distance leading to the spherical kernel density estimate (3.4). The order-restricted estimator $\hat{\boldsymbol{\mu}}^c$ is then defined as

$$\hat{\boldsymbol{\mu}}^c = \mathrm{argmin}_{\boldsymbol{\phi} \in \mathcal{C}} \mathrm{SCE}(\hat{\boldsymbol{\mu}}, \boldsymbol{\phi}). \tag{4.6}$$

By similarity with (4.5), the solution $\hat{\boldsymbol{\mu}}^c$ is termed the *circular isotonic regression estimator (CIRE)*. Applying the PAVA algorithm directly to (4.6) is not possible for two reasons. First, the linear average $(\mu_1 + \mu_2)/2$ has to be replaced with the circular average; see (1.1). Second, since the data lie on the circle, there is no unique violator as in the linear case. Thus, even a circular adaptation of PAVA, which we write CPAVA and which simply replaces the classical average in PAVA with the circular average (1.1), needs to evaluate all the possible settings in terms of SCE, rendering the problem computationally demanding. Rueda et al. (2009) suggested a computationally efficient and exact algorithm to solve (4.6). Their proposal is based on theoretical findings characterizing those situations where the CPAVA yields the CIRE.

This algorithm, and hence the CIRE, is implemented in the R package **isocir** described in Barragán et al. (2013). That package contains a set of functions to analyze angular data under circular order constraints. In particular, it contains a generalization of the CIRE to more general order restrictions of the form

$$\begin{pmatrix} \mu_1 \\ \mu_2 \\ \vdots \\ \mu_i \end{pmatrix} \preceq \begin{pmatrix} \mu_{i+1} \\ \vdots \\ \mu_j \end{pmatrix} \preceq \begin{pmatrix} \mu_{j+1} \\ \vdots \\ \mu_p \end{pmatrix} \tag{4.7}$$

where $1 \leq i < j < p$. In such settings, each element of $\{\mu_1, \ldots, \mu_i\}$ must precede each element of $\{\mu_{i+1}, \ldots, \mu_j\}$, but within each set there are no order restrictions.

4.3 Exploring data features with the CircSiZer

4.3.1 The SiZer, scale space theory and organization of the remainder of the section

In the absence of any parametric assumptions, non-parametric density estimation can provide useful information about the structure of the data. Kernel density estimation was described in Chapter 3. As explained there, the choice of the (here, linear) bandwidth parameter g regulating the smoothness of the estimated curve is a delicate issue. Depending on the value of g, $\hat{f}_g(\cdot)$ from (3.1) can exhibit very diverse shapes, as we illustrate in the left-hand panel of Figure 4.5. A particularly striking feature of this figure is that the number of modes varies with g, prompting the question "Which features are really there?" The seminal papers by Chaudhuri & Marron (1999) and Chaudhuri & Marron (2000) popularized a user-friendly graphical tool, the SiZer (SIgnificant ZERo crossing of derivatives) map, to help answer this question.

The method of Chaudhuri and Marron differs substantially from standard kernel density estimation procedures. Instead of trying to detect the "best" bandwidth g, it uses a variety of distinct bandwidths. Their approach is motivated by ideas from computer vision, namely by scale-space theory. It consists in viewing the

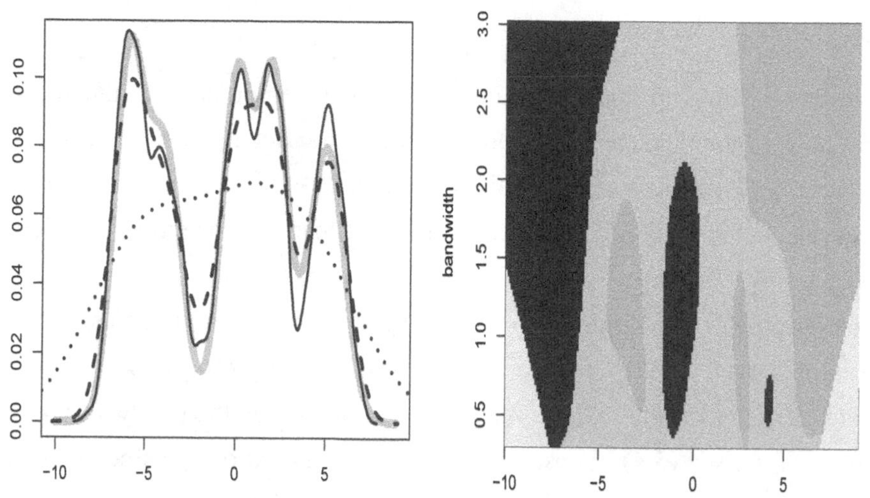

Figure 4.5: The left-hand panel depicts a multimodal mixture of normal densities on the real line (grey) together with kernel density estimates with bandwidth g equal to 0.3 (solid), 0.7 (dashed) and 2.4 (dotted). The right-hand picture exhibits the corresponding SiZer map where the grey level represents the behavior of $\mathrm{E}[\hat{f}_g(z)]$: the areas in black indicate an increasing trend, areas in dark grey a decreasing trend, areas in light grey no significant trend and areas in white a lack of information. Thus, at every fixed value of g one can easily visualize the features of the data over their domain.

same picture at different levels of resolution, hence at different scales. Applied to density estimation, this corresponds to representing various estimates $\hat{f}_g(\cdot)$ at the same time, with large g yielding a macroscopic vision and small g a microscopic vision on the features within the data. A second crucial aspect of this method is a shift of focus from the unknown true density $f(z)$ to $\mathrm{E}[\hat{f}_g(z)]$, function that shares the main features of $f(z)$ for most values of g. Since $\hat{f}_g$ is an unbiased estimator of $\mathrm{E}[\hat{f}_g(z)]$, the bias (3.2), which increases with the bandwidth, is avoided. The goal is hence to extract relevant information about f through differently scaled versions of $\hat{f}_g$. Figure 4.5 gives an idea of how the SiZer works. In order to avoid repetition, we shall not explain its construction here but refer the reader to Section 4.3.2 where we provide details on the CircSiZer.

The remainder of the section is organized as follows. A description of the CircSiZer, related information and illustrations are given in Section 4.3.2, while Section 4.3.3 briefly addresses the issue of the kernel function choice.

4.3.2 The CircSiZer

Oliveira et al. (2014*a*) proposed a circular version of the popular SiZer, namely the CircSiZer. Its implementation in `R` can be found in the package `NPCirc`, described in Oliveira et al. (2014*b*). Besides the CircSiZer, `NPCirc` contains `R` codes for most of the circular and circular-linear density estimation procedures described in Chapter 3.

The purpose of the CircSiZer is to provide a graphical tool to assess the main features of a data set $\Theta_1, \ldots, \Theta_n \in \mathcal{S}^1$ by showing a group of kernel density estimates at distinct scales. These scales are determined by the bandwidth h in the definition of $\hat{f}_h$; see (3.4). Subsequent inference is based on the smoothed curve $f(\theta; h) = \mathrm{E}\left[\hat{f}_h(\theta)\right]$ instead of on $f(\theta)$. It is to be noted that the distance $1 - \mathbf{x}'\mathbf{X}_i$ simplifies in the circular setting to $1 - \cos(\theta - \Theta_i)$ for all $i = 1, \ldots, n$. We shall from here on follow the parsimonious notation of Oliveira et al. (2014*a*) and write $\hat{f}_h(\theta) = \frac{1}{n}\sum_{i=1}^n K_{c;h}(\theta - \Theta_i)$ with circular kernel $K_{c;h}$.

Information about peaks and valleys in the density $f(\theta)$ is contained in the derivative $f'(\theta)$. The CircSiZer, like the SiZer, focusses instead on $f'(\theta; h) = \mathrm{E}\left[\hat{f'}_h(\theta)\right]$ and builds confidence intervals for $f'(\theta; h)$ of the form

$$\left[\hat{f}'(\theta; h) - q_{1-\alpha}\widehat{\mathrm{sd}}(\hat{f}'(\theta; h)); \hat{f}'(\theta; h) - q_{\alpha}\widehat{\mathrm{sd}}(\hat{f}'(\theta; h))\right] \tag{4.8}$$

where q_α and $q_{1-\alpha}$ are the α-lower and upper quantiles of the distribution of $\hat{f}'(\theta; h)$ with mean $f'(\theta; h)$ and standard deviation $\mathrm{sd}(\hat{f}'(\theta; h))$. The latter quantity is estimated in (4.8) as the square root of

$$\widehat{\mathrm{Var}}\left[\hat{f}'(\theta; h)\right] = \frac{1}{n}s^2\left((K_{c;h})'(\theta - \Theta_1), \ldots, (K_{c;h})'(\theta - \Theta_n)\right)$$

where $s^2(\cdot)$ stands for the sample variance of the derivatives of the $K_{c;h}(\theta - \Theta_i)$'s. From the interval (4.8) we readily draw the following conclusions:

(a) The interval (4.8) has a lower value that is positive $\Rightarrow$ increasing trend for $f(\theta; h)$;

(b) The interval (4.8) has an upper value that is negative $\Rightarrow$ decreasing trend for $f(\theta; h)$;

(c) The interval (4.8) contains 0 $\Rightarrow$ no significant trend for $f(\theta; h)$.

For any couple (θ, h), one of these situations will hold. Letting θ vary in $[-\pi, \pi)$ and h between 0 and some upper value $h_{\max}$ leads to a disc with radius $h_{\max}$, disc that can be divided into areas satisfying any of the three conditions. The areas are marked with distinct scales of grey: black for areas of type (a), dark grey for areas of type (b) and grey for areas of type (c). A fourth shade, light grey, is added to indicate those areas where insufficient data is available to draw meaningful conclusions. All couples (θ, h) for which the estimated effective sample size

$$\mathrm{ESS}(\theta, h) = \frac{\sum_{i=1}^{n} K_{c;h}(\theta - \Theta_i)}{K_{c;h}(0)}$$

satisfies $\mathrm{ESS}(\theta, h) < 5$ are considered as too weak in information. Thus, the grey-scale disc consists for each $h \in (0, h_{\max}]$ of a color ring with four colors. This circular color map is the CircSiZer.

It is straightforward to infer significant features from the CircSiZer map. At a fixed bandwidth, a significant peak corresponds to a black region followed by a dark grey region, with potentially a small grey region in between. Conversely, the sequence dark grey to black identifies a valley in the data. We illustrate the *modus operandi* of the CircSiZer by means of two examples in Figure 4.6, where it is applied to one asymmetric unimodal density and one multimodal density.

A technical difficulty in the construction of the CircSiZer is the estimation of the quantiles q_α and $q_{1-\alpha}$. Oliveira et al. (2014*a*) opted for a bootstrap approach, whose mechanism is summarized in the following steps:

1. Generate B random samples drawn with replacement from the original data.

2. For $b = 1, \ldots, B$, compute

$$V_b^* := \frac{\hat{f}'(\theta; h)_b^* - \hat{f}'(\theta; h)}{\widehat{\mathrm{sd}}\left(\hat{f}'(\theta; h)_b^*\right)}$$

where $\hat{f}'(\theta; h)_b^*$ is obtained on basis of the b-th bootstrap sample and $\widehat{\mathrm{sd}}\left(\hat{f}'(\theta; h)_b^*\right)$ estimates its standard deviation.

3. Compute the α-lower and upper quantiles of the series $V_1^*, \ldots, V_B^*$, and denote them $\hat{t}^{(\alpha)}$ and $\hat{t}^{(1-\alpha)}$.

This algorithm, called the "bootstrap-t" approach (hence the notation), yields $q_\alpha = \hat{t}^{(\alpha)}$ and $q_{1-\alpha} = \hat{t}^{(1-\alpha)}$. Note that the bootstrap standard deviations are estimated in the same way as $\widehat{\mathrm{sd}}(\hat{f}'(\theta; h))$.

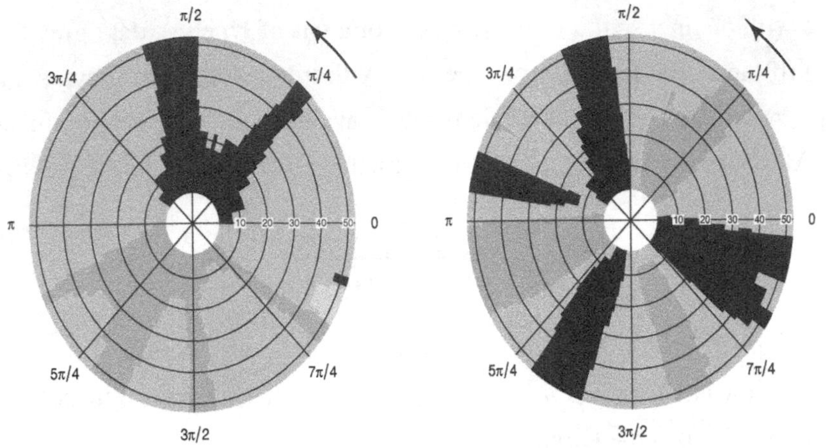

Figure 4.6: Left-hand picture: CircSiZer map for a sine-skewed von Mises density (2.6) with $\mu = \pi, \kappa = 1$ and $\lambda = 0.4$. Right-hand picture: CircSiZer map for a multimodal mixture of von Mises densities with four modes, one near 0, two between $3\pi/4$ and π and one close to $3\pi/2$ (the mixture corresponds to an equal combination of five von Mises densities with the following pairs of parameters (μ, κ): $(0, 8), (\pi/4, 4), (3\pi/4, 10), (\pi, 10)$ and $(3\pi/2, 4)$). In each case, the CircSiZer bandwidths vary between 5 and 50, and the following color legend holds: areas in black stand for an increasing trend, areas in dark grey for a decreasing trend, areas in grey for no significant trend and light grey means a lack of information at the corresponding area.

4.3.3 Kernel choice based on causality: the special role of the wrapped normal

A more formal treatment of the CircSiZer and associated circular scale-space theory was proposed in Huckemann et al. (2016). In particular, they considered the issue of causality, which means that the number of modes should be decreasing when the smoothing increases. In the linear setting, only the normal kernel can assure this property. In the circular case, the situation is similar: the only causal kernel is the wrapped normal kernel. As a reminder, the wrapped normal density corresponds to using $f_Z(z) = (2\pi)^{-1/2} \exp(-z^2/2)$ in (2.1). Therefore, to ensure proper inference, we recommend using this kernel; the von Mises kernel is computationally simpler and also provides useful information about the data.

4.4 Computationally fast estimation for high-dimensional FvML distributions

4.4.1 Maximum likelihood expressions for the parameters of FvML distributions and organization of the section

For the ease of readability, we recall the density of a p-dimensional Fisher–von Mises–Langevin distribution:

$$\mathbf{x} \mapsto \frac{\left(\frac{\kappa}{2}\right)^{p/2-1}}{2\pi^{p/2} I_{p/2-1}(\kappa)} \exp(\kappa \mathbf{x}'\boldsymbol{\mu})$$

where $I_{p/2-1}$ is the modified Bessel function of the first kind and of order $p/2-1$, described in detail in Section 2.2.2. Further details on the FvML distribution are provided in Section 2.3.1. Letting $\mathbf{X}_1, \ldots, \mathbf{X}_n$ be iid observations from an FvML population, it is easy to derive the maximum likelihood estimates for the location parameter $\boldsymbol{\mu} \in \mathcal{S}^{p-1}$ and the concentration parameter $\kappa \geq 0$:

$$\hat{\mu}_{\text{MLE}} = \frac{\bar{\mathbf{X}}}{||\bar{\mathbf{X}}||} \quad \text{and} \quad \hat{\kappa}_{\text{MLE}} = A_p^{-1}(\bar{R}) \tag{4.9}$$

with $\bar{\mathbf{X}} = \frac{1}{n}\sum_{i=1}^{n} \mathbf{X}_i$, $\bar{R} = ||\bar{\mathbf{X}}||$ and

$$A_p(\kappa) := \frac{I_{p/2}(\kappa)}{I_{p/2-1}(\kappa)}. \tag{4.10}$$

Despite being defined through a closed-form expression, the ML estimate of κ involves the inverse of a ratio of modified Bessel functions. Given the complexity of these functions, computational methods are required to achieve this inversion.

The rest of the section is organized as follows. We start by providing the state-of-the-art approximations for $\hat{\kappa}$ from around the year 2000, and discuss their limitations in modern applications (Section 4.4.2). Section 4.4.3 contains various new approximations proposed over the past decade to deal especially with high-dimensional settings, and we provide a numerical comparison between those proposals.

4.4.2 Approximations for the concentration parameter from Mardia & Jupp (2000) and their limitations in high dimensions

In numerous applications involving directional statistics, the data are assumed to follow an FvML distribution. It is hence not surprising that several solutions have been put forward to overcome the computational complexity related to the estimation of κ. Section 10.3.1 of Mardia & Jupp (2000) contains the following two approximations for $\hat{\kappa}_{\mathrm{MLE}}$ (from here on we drop the index MLE):

- when κ is large and $\bar{R}$ close to 1:

$$\hat{\kappa} \approx \frac{p-1}{2(1-\bar{R})}; \tag{4.11}$$

- when κ is small as well as $\bar{R}$:

$$\hat{\kappa} \approx p\bar{R}\left(1 + \frac{p}{p+2}\bar{R}^2 + \frac{p^2(p+8)}{(p+2)^2(p+4)}\bar{R}^4\right). \tag{4.12}$$

These numerical approximations assume either that p is not too large or κ is very small compared to p. Take for example $p = 1000, \kappa = 800$ and $\bar{R} \approx 0.55$: (4.11) and (4.12) respectively give 1120.92 and 776.80 as approximated estimates for κ while for instance the improved formula (4.13) yields 800.13 (these results are taken from Banerjee et al. 2005). Thus, approximations (4.11) and (4.12) are not suited for various situations including high-dimensional datasets. The latter have however become a characteristic of contemporary science and of modern statistical research; see Chapter 7. Moreover, several high-dimensional datasets from text mining (see Section 1.2.3 of the Introduction) or genetics will typically lead to $\bar{R}$ neither too large nor too small. These limitations of the Mardia–Jupp approximations have motivated the research community to work out improved methods to accurately estimate κ, which we shall describe in what follows.

4.4.3 New (high-dimensional) approximations for the concentration parameter

One of the first high-dimensional approximations was proposed in Banerjee et al. (2005). Noting that the ratio of Bessel functions (4.10) allows the continued fraction representation

$$A_p(\kappa) = \cfrac{1}{\frac{p}{\kappa} + \cfrac{1}{\frac{p+2}{\kappa} + \cdots}}$$

and, writing $A_p(\hat{\kappa}) = \bar{R}$, we get the approximation $\frac{1}{\bar{R}} \approx \frac{p}{\hat{\kappa}} + \bar{R}$ and hence $\hat{\kappa} \approx p\bar{R}/(1-\bar{R}^2)$. Banerjee et al. (2005) noticed on an empirical basis that the latter expression can be further improved in the high-dimensional setting and especially for medium-size values of $\bar{R}$ via the correction term $-\bar{R}^3/(1-\bar{R}^2)$, resulting in

$$\hat{\kappa} \approx \frac{\bar{R}p - \bar{R}^3}{1-\bar{R}^2} \quad \text{(Banerjee et al. 2005 approximation)}. \tag{4.13}$$

A different path is taken by Tanabe et al. (2007), who made use of an iterative algorithm based on fixed-point theory. Rewriting (4.9) as $\hat{\kappa} = \hat{\kappa}\bar{R}/A_p(\hat{\kappa})$, consider the recurrence formula

$$\kappa_{r+1} = \frac{\kappa_r \bar{R}}{A_p(\kappa_r)}.$$

If the sequence $(\kappa_r)_{r\in\mathbb{N}}$ converges as $r \to \infty$, then the limit is a fixed point. Tanabe et al. (2007) showed that the mapping $\kappa \mapsto \kappa\bar{R}/A_p(\kappa)$ admits a unique fixed point, hence necessarily $\lim_{r\to\infty} \kappa_r = \hat{\kappa}$ if the limit converges, motivating their iterative procedure. Of course, the entire procedure is speeded up by using a good starting point. To this end, Tanabe et al. (2007) used the Amos-type inequality

$$\frac{1}{\frac{p}{\kappa} + A_p(\kappa)} \leq A_p(\kappa) \leq \frac{\kappa}{p/2 - 1 + \sqrt{\kappa^2 + (p/2-1)^2}}. \tag{4.14}$$

Inequalities of this type and further results on modified Bessel functions and their ratios were studied in the seminal paper by Amos (1974). Straightforward calculations allow deducing from (4.14) that

$$\frac{\bar{R}(p-2)}{1-\bar{R}^2} \leq \hat{\kappa} \leq \frac{\bar{R}p}{1-\bar{R}^2} \qquad \text{(Tanabe et al. 2007 bounds)}. \tag{4.15}$$

The fixed point iterative method thus uses an initial value from this interval. One can further deduce from (4.15) that

$$\hat{\kappa} = \frac{\bar{R}(p-c)}{1-\bar{R}^2} \qquad \text{(Tanabe et al. 2007 formula)} \tag{4.16}$$

for some constant $c \in [0,2]$. Note that (4.13) corresponds to $c = \bar{R}^2$, while $c = 1$ yields the mid-point value. Moreover, writing κ_l and κ_u the lower and upper bounds in (4.15) and defining $\Phi_p(\kappa) = \bar{R}\kappa/A_p(\kappa)$, a linear interpolation argument yields the further approximation

$$\hat{\kappa} = \frac{\kappa_l \Phi_p(\kappa_u) - \kappa_u \Phi_p(\kappa_l)}{\Phi_p(\kappa_u) - \Phi_p(\kappa_l) - (\kappa_u - \kappa_l)} \qquad \text{(Tanabe et al. 2007 approximation)}.$$

Sra (2011) suggested improving (4.13) via two iterations of the Newton method applied on $A_p(\kappa) = \bar{R}$. Starting with κ_0 given by (4.13) and using the formula

$$A_p'(\kappa) = 1 - A_p(\kappa)^2 - \frac{p-1}{\kappa} A_p(\kappa),$$

this approach consists in computing successively

$$\kappa_{N;1} = \kappa_0 - \frac{A_p(\kappa_0) - \bar{R}}{A_p'(\kappa_0)}$$

and

$$\kappa_{N;2} = \kappa_{N;1} - \frac{A_p(\kappa_{N;1}) - \bar{R}}{A_p'(\kappa_{N;1})} \qquad \text{(Sra 2011 approximation)}\,.$$

The restriction to two iterations to approximate $\hat{\kappa}$ via $\kappa_{N;2}$ is motivated by the parsimony of calculations of the ratio $A_p(\kappa)$.

A similar approach is adopted by Song et al. (2012), who however used Halley iterations to solve $A_p(\kappa) = \bar{R}$, based on a second-order Taylor expansion. Thanks to the second derivative formula

$$A_p''(\kappa) = 2A_p(\kappa)^3 + \frac{3(p-1)}{\kappa}A_p(\kappa)^2 + \frac{p^2 - p - 2\kappa^2}{\kappa^2}A_p(\kappa) - \frac{p-1}{\kappa}$$

a Halley-based two-step iteration, starting again from κ_0 as in (4.13), results in

$$\kappa_{H;1} = \kappa_0 - \frac{2\left(A_p(\kappa_0) - \bar{R}\right) A_p'(\kappa_0)}{2(A_p'(\kappa_0))^2 - \left(A_p(\kappa_0) - \bar{R}\right) A_p''(\kappa_0)}$$

and (Song et al. 2012 approximation)

$$\kappa_{H;2} = \kappa_{H;1} - \frac{2\left(A_p(\kappa_{H;1}) - \bar{R}\right) A_p'(\kappa_{H;1})}{2(A_p'(\kappa_{H;1}))^2 - \left(A_p(\kappa_{H;1}) - \bar{R}\right) A_p''(\kappa_{H;1})}.$$

Both $\kappa_{H;2}$ and $\kappa_{N;2}$ only require two calculations of $A_p(\kappa)$.

Hornik & Grün (2014) reinvestigated the upper and lower bounds of Tanabe et al. (2007). Defining the function

$$G_{\alpha,\beta}(\kappa) := \frac{\kappa}{\alpha + \sqrt{\kappa^2 + \beta^2}}$$

they derived the following tighter bounds on $A_p(\kappa)$:

$$G_{p/2-1/2,p/2+1/2}(\kappa) \leq A_p(\kappa) \leq \min\left(G_{p/2-1,p/2+1}(\kappa), G_{p/2-1/2,\sqrt{p^2-1}/2}(\kappa)\right). \tag{4.17}$$

Note how the right-hand side of (4.14) can be expressed as $G_{p/2-1,p/2-1}(\kappa)$, which can be shown to be larger than the right-hand side of (4.17); the left-hand side of (4.17) also improves on that in (4.14). A particularly appealing feature of $G_{\alpha,\beta}$ is that this function is strictly increasing on $[0,\infty)$ and hence allows a unique inverse given by

$$G_{\alpha,\beta}^{-1}(\bar{R}) = \frac{\bar{R}}{1-\bar{R}^2}\left(\alpha + \sqrt{\bar{R}^2\alpha^2 + (1-\bar{R}^2)\beta^2}\right).$$

This readily leads to tighter upper and lower bounds for the maximum likelihood estimate (Hornik & Grün 2014 bounds)

$$\max\left(G^{-1}_{p/2-1,p/2+1}(\bar{R}), G^{-1}_{p/2-1/2,\sqrt{p^2-1}/2}(\bar{R})\right) \leq \hat{\kappa} \leq G^{-1}_{p/2-1/2,p/2+1/2}(\bar{R}).$$

$(p, \kappa_{\text{true}})$	Banerjee et al. (2005)	Tanabe et al. (2007)	Sra (2011)	Song et al. (2012)
$(10, 10)$	1.63E-01	1.50E-02	2.56E-07	6.23E-21
$(10, 100)$	4.54E-01	3.34E-02	4.01E-08	5.28E-33
$(10, 1000)$	4.95E-01	3.43E-02	5.98E-11	9.80E-49
$(100, 10)$	9.26E-04	3.54E-07	1.69E-20	3.44E-44
$(100, 100)$	1.70E-01	1.03E-03	2.64E-10	3.39E-29
$(100, 1000)$	4.52E-01	2.44E-03	3.87E-11	1.11E-40
$(1000, 10)$	9.96E-07	3.98E-13	2.63E-38	9.51E-79
$(1000, 100)$	9.58E-04	3.65E-08	2.03E-23	4.93E-52
$(1000, 1000)$	1.71E-01	9.93E-05	2.64E-13	3.19E-37

Table 4.1: Absolute approximation errors $|\hat{\kappa} - \kappa_{\text{true}}|$ for distinct pairs $(p, \kappa_{\text{true}})$. These values are taken from numerical experiments in Song et al. (2012), who used 1000 digits of precision.

Like (4.15), this interval allows choosing a meaningful initial value for iterative algorithms such as the fixed-point algorithm of Tanabe et al. (2007) or some other root-finding algorithm such as simple bisection.

We conclude this section with some comparative comments on the various aforementioned approximations. First, the tighter Hornik & Grün (2014) bounds obviously lead to a (slightly) quicker convergence of iterative algorithms than the Tanabe et al. (2007) bounds. A comparison of the performance of the four approximations for $\hat{\kappa}$ is shown in Table 4.1 based on numerical experiments conducted in Song et al. (2012). We have chosen the three possible situations, namely p close to κ, $p >> \kappa$ and $p << \kappa$. Unsurprisingly, the most recent Song et al. (2012) approximation does best. However, in very high dimensions computational feasibility is extremely important, and here the Banerjee et al. (2005) approximation is recommended, as it is very simple, yields good results and needs no evaluation of $A_p(\kappa)$. In terms of computational cost, the other three approximations are similar given that they require two evaluations of $A_p(\kappa)$, hence one may want to choose the most accurate, although all three perform very well. As remarked by several authors, the increased accuracy happens to be more of an academic concern.

4.5 Further reading

Ordering data on the sphere: quantiles and depth functions

We defined in Section 4.1.2 three distinct notions of spherical depth functions. For properties of these, such as monotonicity or determination of the maximal value of a depth function, we refer the reader to Liu & Singh (1992) and Agostinelli & Romanazzi (2013). The latter paper also nicely illustrates their uses in practice by analyzing animal orientation and wind data.

Statistical inference under order restrictions on the circle

We provided in Section 4.2 an introduction to order restricted inference for circular data, with particular focus on circular isotonic regression. Testing for a particular order $\mathcal{C}$ among angular parameters was considered in Fernández et al. (2012) and extended to partial orders of the type (4.7) by Barragán et al. (2013). The testing problem may be formulated as follows:

$$\mathcal{H}_0 : \mu_1, \ldots, \mu_p \text{ follow a specific known order}$$
$$\mathcal{H}_1 : \mathcal{H}_0 \text{ is not true.}$$

The **isocir** package includes R codes for the respective testing procedures.

The very recent article by Barragán et al. (2015) addressed the problem of checking whether the temporal order among the components of an oscillatory biological system is unchanged in different populations. This article also provides numerous references to domains where order-restricted inference can play an important role, including cell biology, genetics and evolutionary psychology.

Circular order-restricted inference in psychology is developed in the papers by Klugkist et al. (2012), Baayen et al. (2012) and Baayen & Klugkist (2014). When dealing with directions or measurements on circular scales, psychologists often wish to test their ideas of "natural" orders, which is where the methods and algorithms of this section come into play.

For an extensive overview of the developments in order restricted circular inference, we refer the interested reader to the chapter by Rueda et al. (2015) in the book by Dryden & Kent (2015) written in honor of Professor Kanti Mardia.

Exploring data patterns with the CircSiZer

We described in Section 4.3 the CircSiZer, an exploratory data analysis tool for detecting major patterns in circular data, and briefly mentioned circular scale space issues. We refer the interested reader to Huckemann et al. (2016) for a detailed formal development of this theory (with proofs) as well as for an insightful application to stem cell stress fibre structures. Formal tests for the number of modes in the data $\Theta_1, \ldots, \Theta_n$ were also investigated in Huckemann et al. (2016).

The focus of Section 4.3 lies on circular density features, but the CircSiZer can as well be used for regression models with circular predictors and linear responses. The specificities of the CircSiZer in this setting can be found in Oliveira et al. (2014*a*).

Computationally fast parameter estimation for high-dimensional Fisher–von Mises–Langevin distributions

We retraced in Section 4.4.3 the development of new improved approximations for the maximum likelihood estimate of the concentration parameter in FvML families. We refer the interested reader to the respective papers for more detailed derivations of the individual results. We also point out that Baricz (2014) corrected an error from Tanabe et al. (2007), and that the paper by Banerjee et al. (2005) is backed up by the technical report by Dhillon & Sra (2003).

Yet another article tackling this problem from an alternative angle is provided by Christie (2015). The paper makes use of (4.16) by considering c as of function of $\bar{R}$ (as is the case for (4.13)) and then builds a Taylor series expansion

$$c(\bar{R}) = \sum_{i=0}^{N} \frac{c^{(i)}(\bar{R}_0)}{i!}(\bar{R} - \bar{R}_0)^i$$

where the derivatives $c^{(i)}(\cdot), i = 1, \ldots, N$, are iteratively calculated and where $\bar{R}_0 = A_p(\kappa_0)$ with κ_0 given by (4.13). The advantage of this approach is that A_p is only evaluated once, in the expression for $\bar{R}_0$; the precision of the approximation is of course improved with the size of N. We refer the reader to Christie (2015) for the expressions of the derivatives $c^{(i)}(\cdot)$.

We have focussed so far only on approximations of maximum likelihood estimates for a single FvML distribution in high dimensions. The same issue has also been investigated for mixtures of FvML distributions as well as for Watson distributions, where relevant references are, respectively, Banerjee et al. (2005) and Sra

& Karp (2013). We are not entering into further details here, as these issues will be part of the companion book *Applied Directional Statistics: Modern Methods and Case Studies* (see the Preface to the present book).

Local asymptotic normality for directional data

5.1 Introduction

5.1.1 The LAN property on $\mathbb{R}^p$ and its deep impact on asymptotic statistics

The notion of *local asymptotic normality* of a sequence of statistical models (hereafter also called "experiments") was introduced in the seminal paper "Locally asymptotically normal families of distributions" by Le Cam (1960). That paper figures among the best-known contributions of Lucien Le Cam to mathematical statistics. Part of his work has been presented in the more recent paper by van der Vaart (2002). The Le Cam asymptotic theory of experiments, and in particular the Local Asymptotic Normality (LAN) property, is the backbone of many different recent contributions to the statistical literature. LAN-type results have been established or used in many different statistical contexts on $\mathbb{R}^p$:

- *Time series*: long disturbances (Hallin et al. 1999), efficient estimation in nonlinear autoregressive models (Koul & Schick 1997), multivariate ARMA models (Garel & Hallin 1995, Hallin & Paindaveine 2004), unit root tests (Hallin et al. 2011) and GARCH models (Francq & Zakoïan 2004, 2012).

- *Semiparametric models*: general results (Bickel et al. 1993, Choi et al. 1996, Hallin & Werker 2003), copula models (Genest & Werker 2002, Chen et al. 2006, Segers et al. 2014), inference for elliptical distributions with emphasis on the location parameter (Hallin & Paindaveine 2002), on the scatter parameter (Hallin & Paindaveine 2006) and on Principal Component Analysis (Hallin et al. 2010).

- *Infinite dimensional models*: efficiency in non-parametric models (Begun et al. 1983) and, very recently, inference for high-dimensional models (Onatski et al. 2013, 2014, Cutting et al. 2017*a*).

This selection of references provides an indication of the importance and impact of LAN-based research on $\mathbb{R}^p$, and shows the variety of topics that can be addressed using this methodology.

5.1.2 Organization of the remainder of the chapter

Given the complexity of the topic, we start by providing a detailed description of the fundamentals of the LAN property on $\mathbb{R}^p$. This is achieved in Section 5.2.2. We then explain in Section 5.3.1 how to establish the LAN property for curved experiments, that is, on non-linear manifolds. The targeted extension to hyperspheres is given in Section 5.3.2, while concrete examples of LAN-based inferential procedures for directional data are described in Sections 5.3.3–5.3.5.

5.2 Local asymptotic normality and optimal testing

5.2.1 Contiguity

The concept of *contiguity* was introduced by Le Cam (1960). Let $(\Omega_n, \mathcal{A}_n)$ be measurable spaces equipped with two sequences of probability distributions $\mathrm{P}^{(n)}$ and $\mathrm{Q}^{(n)}$. The sequence $\mathrm{Q}^{(n)}$ is *contiguous* to $\mathrm{P}^{(n)}$ (which we denote as $\mathrm{Q}^{(n)} \lhd \mathrm{P}^{(n)}$) if $\mathrm{P}^{(n)}(A_n) \to 0$ implies $\mathrm{Q}^{(n)}(A_n) \to 0$ for every sequence of measurable sets $A_n \in \mathcal{A}_n$. If $\mathrm{Q}^{(n)} \lhd \mathrm{P}^{(n)}$ and $\mathrm{P}^{(n)} \lhd \mathrm{Q}^{(n)}$, we say that $\mathrm{Q}^{(n)}$ and $\mathrm{P}^{(n)}$ are *mutually contiguous* (and we write $\mathrm{Q}^{(n)} \lhd \rhd \mathrm{P}^{(n)}$). The following famous result is known as the First Le Cam Lemma. It provides some properties for the likelihood ratio $\frac{\mathrm{dP}^{(n)}}{\mathrm{dQ}^{(n)}}$.

Proposition 5.2.1 *(First Le Cam Lemma) Consider two sequences of probability measures* $\mathrm{P}^{(n)}$ *and* $\mathrm{Q}^{(n)}$ *defined on measurable spaces* $(\Omega_n, \mathcal{A}_n)$*. The following are equivalent:*

(i) $\mathrm{Q}^{(n)} \lhd \mathrm{P}^{(n)}$

(ii) $\frac{\mathrm{dP}^{(n)}}{\mathrm{dQ}^{(n)}}$ *converges weakly (under* $\mathrm{Q}^{(n)}$ *) to a random variable* U *along a subsequence, then* $\mathrm{P}(U > 0) = 1$.

(iii) $\frac{\mathrm{dQ}^{(n)}}{\mathrm{dP}^{(n)}}$ *converges weakly (under* $\mathrm{P}^{(n)}$*) to a random variable* V *along a subsequence, then* $\mathrm{E}[V] = 1$.

(iv) Any statistic $S_n : \Omega_n \to \mathbb{R}^p$ *is such that if* $S_n = o_{\mathrm{P}}(1)$ *under* $\mathrm{P}^{(n)}$*, then* $S_n = o_{\mathrm{P}}(1)$ *under* $\mathrm{Q}^{(n)}$.

The First Le Cam Lemma implies that if $\mathrm{P}^{(n)}$ and $\mathrm{Q}^{(n)}$ are sequences such that

$$\frac{\mathrm{dQ}^{(n)}}{\mathrm{dP}^{(n)}} \xrightarrow{\mathcal{D}} \exp(\mathcal{N}(\mu, \sigma^2))$$

under $\mathrm{P}^{(n)}$ as $n \to \infty$ with $\mathcal{N}(\mu, \sigma^2)$ a normal distribution with mean μ and variance σ^2, we have that $\mathrm{P}^{(n)} \lhd \mathrm{Q}^{(n)}$. Furthermore, it can be shown that $\mathrm{P}^{(n)} \lhd\rhd \mathrm{Q}^{(n)}$ if and only if $\mu = -\frac{1}{2}\sigma^2$. The next crucial result is the so-called *Third Le Cam Lemma*. It can be used to compute the asymptotic distribution of a statistic S_n under $\mathrm{Q}^{(n)}$ using the asymptotic distribution of S_n under $\mathrm{P}^{(n)}$, provided that $\mathrm{Q}^{(n)}$ and $\mathrm{P}^{(n)}$ are mutually contiguous.

Proposition 5.2.2 *(Third Le Cam Lemma) Consider two sequences of probability measures* $\mathrm{P}^{(n)}$ *and* $\mathrm{Q}^{(n)}$ *defined on measurable spaces* $(\Omega_n, \mathcal{A}_n)$ *and let* $\mathbf{S}_n : \Omega_n \to \mathbb{R}^p$ *be a statistic. Assume that*

$$\begin{pmatrix} \mathbf{S}_n \\ \log\left(\frac{\mathrm{dQ}^{(n)}}{\mathrm{dP}^{(n)}}\right) \end{pmatrix} \xrightarrow{\mathcal{D}} \mathcal{N}_{p+1}\left(\begin{pmatrix} \boldsymbol{\mu} \\ -\frac{1}{2}\sigma^2 \end{pmatrix}, \begin{pmatrix} \boldsymbol{\Sigma} & \boldsymbol{\tau} \\ \boldsymbol{\tau}' & \sigma^2 \end{pmatrix}\right)$$

under $\mathrm{P}^{(n)}$ *as* $n \to \infty$, *with* $\sigma^2 > 0$, $\boldsymbol{\mu}, \boldsymbol{\tau} \in \mathbb{R}^p$ *and* $\boldsymbol{\Sigma}$ *a* $p \times p$ *real-valued symmetric positive definite matrix. Then,* $\mathbf{S}_n \xrightarrow{\mathcal{D}} \mathcal{N}_p(\boldsymbol{\mu} + \boldsymbol{\tau}, \boldsymbol{\Sigma})$ *under* $\mathrm{Q}^{(n)}$ *as* $n \to \infty$.

The Third Le Cam Lemma is extremely useful for computing the asymptotic power of test procedures under sequences of experiments that are locally and asymptotically normal. A definition of the concept of local asymptotic normality is provided in the following section.

5.2.2 Local asymptotic normality

Let $\mathbf{Z}_1, \ldots, \mathbf{Z}_n$ be an iid sample from a parametric distribution $\mathrm{P}_{\boldsymbol{\vartheta}}$, $\boldsymbol{\vartheta} \in \mathcal{V} \subset \mathbb{R}^p$; write $\mathrm{P}_{\boldsymbol{\vartheta}}^{(n)}$ for the joint distribution of the $\mathbf{Z}_i$'s. Conceptually, a sequence of statistical experiments is locally and asymptotically normal (LAN) if the related sequence of local log-likelihood ratios resembles, asymptotically, those of a Gaussian location model. The following definition of LAN is given in Le Cam & Yang (2000). For all sample size n, let

$$\mathcal{E}^{(n)} = \left(\mathcal{X}^{(n)}, \mathcal{A}^{(n)}, \mathcal{P}^{(n)} := \{\mathrm{P}_{\boldsymbol{\vartheta}}^{(n)} \,|\, \boldsymbol{\vartheta} \in \mathcal{V} \subset \mathbb{R}^p\}\right)$$

be a sequence of $\boldsymbol{\vartheta}$-parametric experiments associated with $\mathbf{Z}_1, \ldots, \mathbf{Z}_n$. The family of probability distributions $\mathcal{P}^{(n)}$ is said to be LAN if for all $\boldsymbol{\vartheta} \in \mathcal{V}$ there exist

(i) a sequence $\boldsymbol{\nu}^{(n)}$ of $p \times p$ full rank non-random matrices such that $\|\boldsymbol{\nu}^{(n)}\| \to 0$ as $n \to \infty$, with $\|\boldsymbol{\nu}^{(n)}\|$ denoting the Frobenius norm of $\boldsymbol{\nu}^{(n)}$;

(ii) a sequence of random p-vectors $\boldsymbol{\Delta}^{(n)}(\boldsymbol{\vartheta})$ called the *central sequence*;

(iii) a non-random $p \times p$ matrix $\boldsymbol{\Gamma}(\boldsymbol{\vartheta})$, the *Fisher information matrix*,

such that for every bounded sequence of vectors $\boldsymbol{\tau}^{(n)} \in \mathbb{R}^p$, we have

$$\log \frac{\mathrm{dP}^{(n)}_{\boldsymbol{\vartheta}+\boldsymbol{\nu}^{(n)}\boldsymbol{\tau}^{(n)}}}{\mathrm{dP}^{(n)}_{\boldsymbol{\vartheta}}} = (\boldsymbol{\tau}^{(n)})'\boldsymbol{\Delta}^{(n)}(\boldsymbol{\vartheta}) - \frac{1}{2}(\boldsymbol{\tau}^{(n)})'\boldsymbol{\Gamma}(\boldsymbol{\vartheta})\boldsymbol{\tau}^{(n)} + o_{\mathrm{P}}(1) \tag{5.1}$$

and $\boldsymbol{\Delta}^{(n)}(\boldsymbol{\vartheta}) \xrightarrow{\mathcal{D}} \mathcal{N}_p(0, \boldsymbol{\Gamma}(\boldsymbol{\vartheta}))$, both under $\mathrm{P}^{(n)}_{\boldsymbol{\vartheta}}$ as $n \to \infty$. Here $\boldsymbol{\nu}^{(n)}$ is called the contiguity rate between the sequences of models $\mathrm{P}^{(n)}_{\boldsymbol{\vartheta}+\boldsymbol{\nu}^{(n)}\boldsymbol{\tau}^{(n)}}$ and $\mathrm{P}^{(n)}_{\boldsymbol{\vartheta}}$.

Example. *We illustrate here the notion of local asymptotic normality on the well-known general linear model. Assume that a sequence $Z_1, \ldots, Z_n$ satisfies*

$$Z_i = \mathbf{Y}_i'\boldsymbol{\beta} + \epsilon_i, \tag{5.2}$$

where $\boldsymbol{\beta} := (\beta_1, \ldots, \beta_p)'$ is the regression parameter, the $\mathbf{Y}_i := (Y_{i1}, \ldots, Y_{ip})'$, $i = 1, \ldots, n$, are the regression constants and where $\epsilon_1, \ldots, \epsilon_n$ is an iid sequence with a strictly positive density f such that

(A1)

$$\int_{\mathbb{R}} z f(z)\, dz = 0; \quad \int_{\mathbb{R}} z^2 f(z)\, dz =: \sigma_f^2 < \infty;$$

(A2) f is absolutely continuous with a.e. derivative $\dot{f}$;

(A3) letting $\varphi_f := -\dot{f}/f$ be the score function,

$$\int_{\mathbb{R}} \varphi_f^2(z) f(z)\, dz =: \mathcal{I}_f < \infty.$$

Letting $\bar{Y}_j := n^{-1}\sum_{i=1}^n Y_{ij}$ and $\mathbf{C}^{(n)} := n^{-1}\sum_{i=1}^n \mathbf{Y}_i\mathbf{Y}_i'$, we also assume that

(A4) $\mathbf{C}^{(n)}$ is positive definite and converges to some positive definite matrix $\mathbf{C}$;

(A5)

$$\frac{\max_{1 \le i \le n}(Y_{ij} - \bar{Y}_j)^2}{\sum_{i=1}^n (Y_{ij} - \bar{Y}_j)^2}$$

is $o(1)$ as $n \to \infty$.

For the general linear model in (5.2), the likelihood of the residuals $Z_i^{\tau}(\boldsymbol{\beta}) := Z_i - \mathbf{Y}_i'\boldsymbol{\beta}$ *is given by* $L^{(n)}(\boldsymbol{\beta}) := \prod_{i=1}^{n} f(Z_i^{\tau}(\boldsymbol{\beta}))$. *Under Assumptions (A1)–(A5), taking* $\boldsymbol{\nu}^{(n)} = n^{-1/2}(\mathbf{C}^{(n)})^{-1/2}$, *we have for any* $\boldsymbol{\beta} \in \mathbb{R}^p$ *and any bounded sequence* $\boldsymbol{\tau}^{(n)}$ *that*

$$\begin{aligned}\Lambda^{(n)}_{f;\boldsymbol{\beta}+\boldsymbol{\nu}^{(n)}\boldsymbol{\tau}^{(n)}/\boldsymbol{\beta}} &:= \log\left(\frac{L^{(n)}(\boldsymbol{\beta}+\boldsymbol{\nu}^{(n)}\boldsymbol{\tau}^{(n)})}{L^{(n)}(\boldsymbol{\beta})}\right)\\ &= (\boldsymbol{\tau}^{(n)})'\boldsymbol{\Delta}_f^{(n)}(\boldsymbol{\beta}) - \frac{1}{2}(\boldsymbol{\tau}^{(n)})'\boldsymbol{\Gamma}_f\boldsymbol{\tau}^{(n)} + o_{\mathrm{P}}(1)\end{aligned}$$

as $n \to \infty$ *under* $\mathrm{P}^{(n)}_{f;\boldsymbol{\beta}}$, *the joint distribution of the* $Z_i^{\tau}(\boldsymbol{\beta})$*'s (equivalently, of the* ϵ_i*'s in (5.2)), where*

$$\boldsymbol{\Delta}_f^{(n)}(\boldsymbol{\beta}) := n^{-1/2}(\mathbf{C}^{(n)})^{-1/2}\sum_{i=1}^{n}\varphi_f(Z_i^{\tau}(\boldsymbol{\beta}))\mathbf{Y}_i$$

and $\boldsymbol{\Gamma}_f = \mathcal{I}_f\mathbf{I}_p$. *It follows from this LAN property that for any statistic* $\mathbf{T}^{(n)}$ *such that* $((\mathbf{T}^{(n)})', (\boldsymbol{\Delta}^{(n)}(\boldsymbol{\beta}))')'$ *is asymptotically normal under* $\mathrm{P}^{(n)}_{f;\boldsymbol{\beta}}$, *one can easily derive the asymptotic behavior of* $\mathbf{T}^{(n)}$ *under local alternatives* $\mathrm{P}^{(n)}_{f;\boldsymbol{\beta}+\boldsymbol{\nu}^{(n)}\boldsymbol{\tau}^{(n)}}$ *using Proposition 5.2.2.*

We easily see that the log-likelihood ratio $\log \frac{d\mathrm{P}^{(n)}_{\boldsymbol{\vartheta}+\boldsymbol{\nu}^{(n)}\boldsymbol{\tau}^{(n)}}}{d\mathrm{P}^{(n)}_{\boldsymbol{\vartheta}}}$ in (5.1) behaves asymptotically like the log-likelihood ratio of the classical Gaussian shift experiment

$$\mathcal{E}_{\boldsymbol{\Gamma}(\boldsymbol{\vartheta})} = (\mathbb{R}^p, \mathcal{B}_p, \mathcal{P}_{\boldsymbol{\vartheta}} := \{\mathcal{N}_p(\boldsymbol{\Gamma}(\boldsymbol{\vartheta})\boldsymbol{\tau}, \boldsymbol{\Gamma}(\boldsymbol{\vartheta})) | \boldsymbol{\tau} \in \mathbb{R}^p\}),$$

with a single observation which we denote as $\boldsymbol{\Delta}$. Here, $\mathcal{B}_p$ denotes the Borel sigma-field on $\mathbb{R}^p$. This approximation of the statistical experiments $\mathcal{E}^{(n)}$ by the normal experiment $\mathcal{E}_{\boldsymbol{\Gamma}(\boldsymbol{\vartheta})}$ has important consequences for the construction of locally and asymptotically optimal test procedures as it means that, asymptotically, all power functions that are implementable in the local experiments $\mathcal{E}^{(n)}$ are the power functions that are possible in the limiting Gaussian shift experiment $\mathcal{E}_{\boldsymbol{\Gamma}(\boldsymbol{\vartheta})}$. In view of these considerations, it follows that asymptotically optimal tests in the local models can be derived by analyzing the Gaussian limit model. More precisely, if a test $\phi(\boldsymbol{\Delta})$ enjoys some exact optimality property in the Gaussian experiment $\mathcal{E}_{\boldsymbol{\Gamma}(\boldsymbol{\vartheta})}$, then the corresponding sequence $\phi(\boldsymbol{\Delta}^{(n)})$ inherits, locally and asymptotically, the same optimality properties in the sequence of experiments $\mathcal{E}^{(n)}$.

Now, let $f_{\boldsymbol{\vartheta}}$ be a density of $\mathrm{P}_{\boldsymbol{\vartheta}}$ with respect to some measure m. It can be shown that the LAN property (5.1) is essentially a consequence of a condition that involves the first derivative of $f_{\boldsymbol{\vartheta}}^{1/2}$: the quadratic mean differentiability of $\boldsymbol{\vartheta} \mapsto f_{\boldsymbol{\vartheta}}^{1/2}$ (see Theorem 7.2 in van der Vaart 1998). The mapping $\boldsymbol{\vartheta} \mapsto f_{\boldsymbol{\vartheta}}^{1/2}$ is differentiable in quadratic mean if there exists a function $\dot{f}_{\boldsymbol{\vartheta}}$ such that

$$\int \left(f_{\boldsymbol{\vartheta}+\boldsymbol{\tau}}^{1/2} - f_{\boldsymbol{\vartheta}}^{1/2} - \frac{1}{2}\boldsymbol{\tau}' \dot{f}_{\boldsymbol{\vartheta}} f_{\boldsymbol{\vartheta}}^{-1/2} \right)^2 \mathrm{dm}$$

is $o(\|\boldsymbol{\tau}\|^2)$ as $\|\boldsymbol{\tau}\| \to 0$.

A slightly reinforced version of LAN is ULAN, the *Uniform Local Asymptotic Normality*. Using the same notations as for LAN, a sequence of experiments is said to be ULAN if for every $\boldsymbol{\vartheta} \in \mathcal{V}$, any sequence $\boldsymbol{\vartheta}^{(n)}$ of the form $\boldsymbol{\vartheta}^{(n)} = \boldsymbol{\vartheta} + O(\|\boldsymbol{\nu}^{(n)}\|)$ and any bounded sequence $\boldsymbol{\tau}^{(n)}$, we have

$$\log \frac{\mathrm{dP}^{(n)}_{\boldsymbol{\vartheta}^{(n)}+\boldsymbol{\nu}^{(n)}\boldsymbol{\tau}^{(n)}}}{\mathrm{dP}^{(n)}_{\boldsymbol{\vartheta}^{(n)}}} = (\boldsymbol{\tau}^{(n)})'\boldsymbol{\Delta}^{(n)}(\boldsymbol{\vartheta}^{(n)}) - \frac{1}{2}(\boldsymbol{\tau}^{(n)})'\boldsymbol{\Gamma}(\boldsymbol{\vartheta})\boldsymbol{\tau}^{(n)} + o_{\mathrm{P}}(1) \qquad (5.3)$$

under $\mathrm{P}^{(n)}_{\boldsymbol{\vartheta}}$ as $n \to \infty$. In plain words, the LAN property holds uniformly in a neighborhood of the parameter of interest $\boldsymbol{\vartheta}$. The main attractive feature of ULAN compared to LAN is that it implies the *asymptotic linearity property*

$$\boldsymbol{\Delta}^{(n)}_{\boldsymbol{\vartheta}+\boldsymbol{\nu}^{(n)}\boldsymbol{\tau}^{(n)}} - \boldsymbol{\Delta}^{(n)}_{\boldsymbol{\vartheta}} = \boldsymbol{\Gamma}(\boldsymbol{\vartheta})\boldsymbol{\tau}^{(n)} + o_{\mathrm{P}}(1) \qquad (5.4)$$

under $\mathrm{P}^{(n)}_{\boldsymbol{\vartheta}}$ as $n \to \infty$. This asymptotic linearity plays a fundamental role when dealing with inferential procedures involving statistics based on the central sequence $\boldsymbol{\Delta}^{(n)}_{\boldsymbol{\vartheta}}$. The asymptotic linearity property greatly simplifies the replacement of the unknown parameter $\boldsymbol{\vartheta}$ (or only of the nuisance part of $\boldsymbol{\vartheta}$) in such statistics with *any* root-n consistent estimator satisfying some mild regularity conditions. The control on the asymptotic impact of such a replacement is often obtained by replacing adequately in (5.4) the non-random sequence $\boldsymbol{\tau}^{(n)}$ by the random sequence $(\boldsymbol{\nu}^{(n)})^{-1}(\hat{\boldsymbol{\vartheta}}^{(n)} - \boldsymbol{\vartheta})$ for $\hat{\boldsymbol{\vartheta}}^{(n)}$ such a root-n consistent estimator of $\boldsymbol{\vartheta}$ under $\mathrm{P}^{(n)}_{\boldsymbol{\vartheta}}$.

5.2.3 Optimal testing in LAN experiments

Consider a sequence of univariate experiments

$$\mathcal{E}^{(n)} = \left(\mathcal{X}^{(n)}, \mathcal{A}^{(n)}, \mathcal{P}^{(n)} := \{\mathrm{P}^{(n)}_{\vartheta} | \vartheta \in \mathcal{V} \subset \mathbb{R}\} \right)$$

which is LAN with central sequence $\Delta^{(n)}(\vartheta)$ and contiguity rate $\nu^{(n)} = n^{-1/2}$. A combination of the definition of a LAN sequence with the Third Le Cam Lemma implies that $\Delta^{(n)}(\vartheta)$ converges in distribution to a Gaussian random variable with mean $\Gamma(\vartheta)\tau$ and variance $\Gamma(\vartheta)$ under $\mathrm{P}^{(n)}_{\vartheta+n^{-1/2}\tau}$ as $n \to \infty$ (use $\Delta^{(n)}(\vartheta)$ as statistic S_n in Proposition 5.2.2). Within such a sequence of models, consider the problem of testing

$$\begin{cases} \mathcal{H}_0 : \vartheta \leq \vartheta_0 \\ \mathcal{H}_1 : \vartheta > \vartheta_0 \end{cases} \tag{5.5}$$

for some fixed $\vartheta_0 \in \mathcal{V}$. Taking ϑ of the form $\vartheta = \vartheta_0 + n^{-1/2}\tau$, (5.5) can be rewritten (locally) as

$$\begin{cases} \mathcal{H}_0^{\text{loc}} : \tau \leq 0 \\ \mathcal{H}_1^{\text{loc}} : \tau > 0. \end{cases} \tag{5.6}$$

As explained in the previous section, LAN implies that the sequence of experiments $\mathcal{E}^{(n)}$ converges to the Gaussian shift experiment

$$\mathcal{E}_{\Gamma(\vartheta)} = \left(\mathbb{R}, \mathcal{B}, \mathcal{P}_\vartheta := \{\mathcal{N}\left(\Gamma(\vartheta)\tau, \Gamma(\vartheta)\right) | \, \tau \in \mathbb{R}\}\right).$$

Letting Δ stand for an observation generated from this Gaussian experiment, a uniformly most powerful test for (5.6) rejects the null hypothesis $\mathcal{H}_0^{\text{loc}}$ at the nominal level α when

$$\Gamma(\vartheta_0)^{-1/2}\Delta > z_{1-\alpha},$$

where z_β is the β quantile of the standard normal distribution. It follows from LAN that the test $\phi^{(n)}$ rejecting the null hypothesis $\mathcal{H}_0$ when

$$\Gamma(\vartheta_0)^{-1/2}\Delta^{(n)}(\vartheta_0) > z_{1-\alpha}$$

is locally and asymptotically most powerful within the sequence of experiments $\mathcal{E}^{(n)}$ at asymptotic level α.

Consider now the same problem as in (5.5) except that $\boldsymbol{\tau}$ is multivariate of dimension p. The problem

$$\begin{cases} \mathcal{H}_{0;p} : \boldsymbol{\vartheta} = \boldsymbol{\vartheta}_0 \\ \mathcal{H}_{1;p} : \boldsymbol{\vartheta} \neq \boldsymbol{\vartheta}_0, \end{cases} \tag{5.7}$$

which can be rewritten locally as

$$\begin{cases} \mathcal{H}_{0;p}^{\text{loc}} : \boldsymbol{\tau} = 0 \\ \mathcal{H}_{1;p}^{\text{loc}} : \boldsymbol{\tau} \neq 0, \end{cases} \tag{5.8}$$

is intrinsically multisided. A most powerful test for this problem does not in general exist in Gaussian shift experiments. However, optimal procedures can be obtained by considering so-called *maximin* tests. In all generality, a test ϕ^* is called maximin in the class $\mathcal{C}_\alpha$ of level-α tests for some null hypothesis $\mathcal{H}_0$ against the alternative $\mathcal{H}_1$ if (i) ϕ^* has level α and (ii) the power of ϕ^* is such that

$$\inf_{\mathrm{P}\in\mathcal{H}_1} \mathrm{E}_{\mathrm{P}}[\phi^*] \geq \sup_{\phi\in\mathcal{C}_\alpha} \inf_{\mathrm{P}\in\mathcal{H}_1} \mathrm{E}_{\mathrm{P}}[\phi].$$

In the Gaussian shift model $\mathcal{E}_{\boldsymbol{\Gamma}(\boldsymbol{\vartheta})} = (\mathbb{R}^p, \mathcal{B}_p, \mathcal{P}_{\boldsymbol{\vartheta}} := \{\mathcal{N}_p(\boldsymbol{\Gamma}(\boldsymbol{\vartheta})\boldsymbol{\tau}, \boldsymbol{\Gamma}(\boldsymbol{\vartheta})) | \boldsymbol{\tau} \in \mathbb{R}^p\})$ a maximin test for (5.8) does not exist. However, letting $\boldsymbol{\Delta}$ be an observation from $\mathcal{E}_{\boldsymbol{\Gamma}(\boldsymbol{\vartheta})}$, the test rejecting the null when

$$\boldsymbol{\Delta}'\boldsymbol{\Gamma}^{-1}(\boldsymbol{\vartheta}_0)\boldsymbol{\Delta}$$

exceeds $\chi^2_{p;1-\alpha}$, the α-upper quantile of the chi-squared distribution with p degrees of freedom, is maximin within the class of level-α tests for

$$\begin{cases} \mathcal{H}_{0;p}^{\mathrm{locbis}} : \boldsymbol{\tau} = 0 \\ \mathcal{H}_{1;p}^{\mathrm{locbis}} : \boldsymbol{\tau}'\boldsymbol{\Gamma}(\boldsymbol{\vartheta}_0)\boldsymbol{\tau} > c \end{cases} \tag{5.9}$$

irrespective of $c > 0$. It follows that if a sequence of experiments

$$\mathcal{E}^{(n)} = \left(\mathcal{X}^{(n)}, \mathcal{A}^{(n)}, \mathcal{P}^{(n)} := \{\mathrm{P}_{\boldsymbol{\vartheta}}^{(n)} | \boldsymbol{\vartheta} \in \mathcal{V} \subset \mathbb{R}^p\}\right)$$

is LAN with central sequence $\boldsymbol{\Delta}^{(n)}(\boldsymbol{\vartheta})$, the sequence of tests rejecting the null (at the asymptotic level α) when

$$\boldsymbol{\Delta}^{(n)}(\boldsymbol{\vartheta}_0)'\boldsymbol{\Gamma}^{-1}(\boldsymbol{\vartheta}_0)\boldsymbol{\Delta}^{(n)}(\boldsymbol{\vartheta}_0) > \chi^2_{p;1-\alpha}$$

is locally and asymptotically maximin for the same problem.

Example (Continued). *The problem of testing $\mathcal{H}_0 : \boldsymbol{\beta} = \boldsymbol{\beta}_0$ for some $\boldsymbol{\beta}_0 \in \mathbb{R}^p$ in the general linear model* (5.2) *is clearly of interest. Considering local perturbations of $\boldsymbol{\beta}_0$ of the form $\boldsymbol{\beta}_0 + \boldsymbol{\nu}^{(n)}\boldsymbol{\tau}^{(n)}$, the null hypothesis rewrites $\mathcal{H}_0 : \boldsymbol{\tau}^{(n)} = 0$ in the vicinity of $\boldsymbol{\beta}_0$. The test that rejects the null hypothesis (at the asymptotic level α) when*

$$\mathcal{I}_f^{-1}\|\boldsymbol{\Delta}_f^{(n)}(\boldsymbol{\beta}_0)\|^2 > \chi^2_{p;1-\alpha}$$

is locally and asymptotically maximin in the f-parametric model.

Finally, letting $\mathcal{M}(\mathbf{A})$ stand for the vector space spanned by the columns of a matrix $\mathbf{A}$, consider the problem of testing

$$\begin{cases} \mathcal{H}_0 : \boldsymbol{\tau} \in \mathcal{M}(\boldsymbol{\Upsilon}) \\ \mathcal{H}_1 : \boldsymbol{\tau} \notin \mathcal{M}(\boldsymbol{\Upsilon}), \end{cases} \tag{5.10}$$

where $\boldsymbol{\Upsilon}$ is a $p \times r$ matrix of rank $r < p$. Again, in the Gaussian shift model $\mathcal{E}_{\boldsymbol{\Gamma}(\boldsymbol{\vartheta})}$, a most powerful test for this problem does not exist but a solution can be obtained using the concept of *stringency*. A test ϕ^* is called most stringent in the class of level-α tests $\mathcal{C}_\alpha$ for testing $\mathcal{H}_0$ against $\mathcal{H}_1$ if (i) ϕ^* has level α and (ii) is such that

$$\sup_{\mathrm{P}\in\mathcal{H}_1} r_{\phi^*}(\mathrm{P}) \leq \sup_{\mathrm{P}\in\mathcal{H}_1} r_{\phi}(\mathrm{P}) \quad \forall \phi \in \mathcal{C}_\alpha,$$

where $r_{\phi_0}(\mathrm{P})$ stands for the *regret* of the test ϕ_0 under $\mathrm{P} \in \mathcal{H}_1$ defined as

$$r_{\phi_0}(\mathrm{P}) := \left[\sup_{\phi\in\mathcal{C}_\alpha} \mathrm{E}_\mathrm{P}[\phi]\right] - \mathrm{E}_\mathrm{P}[\phi_0].$$

Thus the regret is the deficiency in power of ϕ_0 under $\mathrm{P} \in \mathcal{H}_1$ compared to the highest possible power under P. Defining $\mathrm{proj}(\mathbf{A}) := \mathbf{A}(\mathbf{A}'\mathbf{A})^{-1}\mathbf{A}'$, the most stringent test within the class of level-α tests for (5.10) in the Gaussian shift model rejects the null when

$$\left\| \left(\mathbf{I}_p - \mathrm{proj}(\boldsymbol{\Gamma}^{1/2}(\boldsymbol{\vartheta})\boldsymbol{\Upsilon})\right) \boldsymbol{\Gamma}^{-1/2}(\boldsymbol{\vartheta})\boldsymbol{\Delta} \right\|^2 > \chi^2_{p-r;1-\alpha}$$

so that in a LAN sequence with central sequence $\boldsymbol{\Delta}^{(n)}(\boldsymbol{\vartheta})$ and contiguity rate $\boldsymbol{\nu}^{(n)}$, the test based on the rejection rule

$$\left\| \left(\mathbf{I}_p - \mathrm{proj}(\boldsymbol{\Gamma}^{1/2}(\boldsymbol{\vartheta})(\boldsymbol{\nu}^{(n)})^{-1}\boldsymbol{\Upsilon})\right) \boldsymbol{\Gamma}^{-1/2}(\boldsymbol{\vartheta})\boldsymbol{\Delta}^{(n)}(\boldsymbol{\vartheta}) \right\|^2 > \chi^2_{p-r;1-\alpha}$$

is locally and asymptotically most stringent.

5.2.4 LAN, semiparametric efficiency and invariance

In this section we consider semiparametric models, that is, models indexed by a finite-dimensional parameter $\boldsymbol{\vartheta} \in \mathcal{V} \subset \mathbb{R}^p$ along with an infinite-dimensional parameter f that belongs to some class of functions $\mathcal{F}$. That is, we consider sequences of experiments of the form

$$\mathcal{E}^{(n)} = \left(\mathcal{X}^{(n)}, \mathcal{A}^{(n)}, \mathcal{P}^{(n)} := \{\mathrm{P}^{(n)}_{\boldsymbol{\vartheta},f} | \boldsymbol{\vartheta} \in \mathcal{V} \subset \mathbb{R}^p, f \in \mathcal{F}\}\right), \tag{5.11}$$

where the parameter f plays the role of a nuisance parameter throughout this section. We first consider a sequence of parametric models

$$\mathcal{P}^{(n)} = \{\mathrm{P}^{(n)}_{\boldsymbol{\vartheta}}, \boldsymbol{\vartheta} := (\boldsymbol{\eta}_1', \boldsymbol{\eta}_2')'; \boldsymbol{\eta}_1 \in \mathbf{N} \subset \mathbb{R}^m, \boldsymbol{\eta}_2 \in \mathbf{H} \subset \mathbb{R}^{m-p}\}$$

such that the resulting sequence of experiments is LAN with central sequence $\left((\boldsymbol{\Delta}_1^{(n)}(\boldsymbol{\vartheta}))', (\boldsymbol{\Delta}_2^{(n)}(\boldsymbol{\vartheta}))'\right)'$ and full rank Fisher information matrix

$$\boldsymbol{\Gamma}(\boldsymbol{\vartheta}) := \begin{pmatrix} \boldsymbol{\Gamma}_{11}(\boldsymbol{\vartheta}) & \boldsymbol{\Gamma}_{12}(\boldsymbol{\vartheta}) \\ \boldsymbol{\Gamma}_{21}(\boldsymbol{\vartheta}) & \boldsymbol{\Gamma}_{22}(\boldsymbol{\vartheta}) \end{pmatrix},$$

where $\boldsymbol{\Delta}_1^{(n)}(\boldsymbol{\vartheta})$ is the $\boldsymbol{\eta}_1$-part and $\boldsymbol{\Delta}_2^{(n)}(\boldsymbol{\vartheta})$ the $\boldsymbol{\eta}_2$-part of the central sequence. Assume now that we are interested in performing inference about $\boldsymbol{\eta}_1$. Intuitively, this is more complicated in the sequence of models $\mathcal{P}^{(n)}$ than in a sequence of submodels of the form

$$\mathcal{P}_{\text{sub}}^{(n)} = \{\mathrm{P}_{\boldsymbol{\eta}_1,\boldsymbol{\eta}_0}^{(n)}; \boldsymbol{\eta}_1 \in \mathbf{N} \subset \mathbb{R}^m\}$$

where $\boldsymbol{\eta}_2$ is fixed as $\boldsymbol{\eta}_0$ (treated as known). The information bound for estimating $\boldsymbol{\eta}_1$ in $\mathcal{P}_{\text{sub}}^{(n)}$ is $\boldsymbol{\Gamma}_{11}^{-1}(\boldsymbol{\vartheta})$, whereas this bound becomes $(\boldsymbol{\Gamma}^{-1}(\boldsymbol{\vartheta}))_{11} = (\boldsymbol{\Gamma}_{11}(\boldsymbol{\vartheta}) - \boldsymbol{\Gamma}_{12}(\boldsymbol{\vartheta})\boldsymbol{\Gamma}_{22}^{-1}(\boldsymbol{\vartheta})\boldsymbol{\Gamma}_{21}(\boldsymbol{\vartheta}))^{-1}$ in $\mathcal{P}^{(n)}$. Such an information bound can be attained by considering asymptotic inference based on the *parametric efficient central sequence*

$$\tilde{\boldsymbol{\Delta}}_1^{(n)}(\boldsymbol{\vartheta}) := \boldsymbol{\Delta}_1^{(n)}(\boldsymbol{\vartheta}) - \boldsymbol{\Gamma}_{12}(\boldsymbol{\vartheta})\boldsymbol{\Gamma}_{22}^{-1}(\boldsymbol{\vartheta})\boldsymbol{\Delta}_2^{(n)}(\boldsymbol{\vartheta}). \tag{5.12}$$

As a consequence, working with unspecified $\boldsymbol{\eta}_2$ decreases the information on $\boldsymbol{\eta}_1$ by $\boldsymbol{\Gamma}_{12}(\boldsymbol{\vartheta})\boldsymbol{\Gamma}_{22}^{-1}(\boldsymbol{\vartheta})\boldsymbol{\Gamma}_{21}(\boldsymbol{\vartheta})$. Note that if $\boldsymbol{\Gamma}_{12}(\boldsymbol{\vartheta}) = \mathbf{0}$ it is possible to perform inference equally well on $\boldsymbol{\eta}_1$ in $\mathcal{P}^{(n)}$ and $\mathcal{P}_{\text{sub}}^{(n)}$. The condition $\boldsymbol{\Gamma}_{12}(\boldsymbol{\vartheta}) = \mathbf{0}$ is a necessary condition for *adaptivity* in such parametric models (see Chen & Bickel 2006 for details). The central sequence $\boldsymbol{\Delta}_1^{(n)}(\boldsymbol{\vartheta})$ and its efficient version $\tilde{\boldsymbol{\Delta}}_1^{(n)}(\boldsymbol{\vartheta})$ can be interpreted geometrically in the Hilbert space of $\mathrm{P}_{\boldsymbol{\vartheta}}^{(n)}$-square-integrable functions. More precisely, $\tilde{\boldsymbol{\Delta}}_1^{(n)}(\boldsymbol{\vartheta})$ can be seen as the projection of $\boldsymbol{\Delta}_1^{(n)}(\boldsymbol{\vartheta})$ on the orthogonal complement of $\boldsymbol{\Delta}_2^{(n)}(\boldsymbol{\vartheta})$ in $L^2(\mathrm{P}_{\boldsymbol{\vartheta}}^{(n)})$, which we denote by

$$\tilde{\boldsymbol{\Delta}}_1^{(n)}(\boldsymbol{\vartheta}) := Pr[\boldsymbol{\Delta}_1^{(n)}(\boldsymbol{\vartheta})|(\boldsymbol{\Delta}_2^{(n)}(\boldsymbol{\vartheta}))^{\perp}].$$

This projection takes into account the information loss due to not knowing $\boldsymbol{\eta}_2$, and the new central sequence (5.12) enjoys efficiency features under these circumstances.

In a similar fashion, the nuisance parameter f in the semiparametric experiments (5.11) renders asymptotic inference on $\boldsymbol{\vartheta}$ more difficult than in fixed-f parametric models. Semiparametric efficiency has to be understood in the Le Cam sense as follows. Assume as previously that the sequence of f-parametric models is LAN, meaning that as $n \to \infty$

$$\log \frac{dP^{(n)}_{\boldsymbol{\vartheta}+\boldsymbol{\nu}^{(n)}\boldsymbol{\tau}^{(n)},f}}{dP^{(n)}_{\boldsymbol{\vartheta},f}} = (\boldsymbol{\tau}^{(n)})'\boldsymbol{\Delta}^{(n)}_f(\boldsymbol{\vartheta}) - \frac{1}{2}(\boldsymbol{\tau}^{(n)})'\boldsymbol{\Gamma}_f(\boldsymbol{\vartheta})\boldsymbol{\tau}^{(n)} + o_P(1) \qquad (5.13)$$

and $\boldsymbol{\Delta}^{(n)}_f(\boldsymbol{\vartheta})$ is asymptotically normal with mean zero and covariance $\boldsymbol{\Gamma}_f(\boldsymbol{\vartheta})$, both under $P^{(n)}_{\boldsymbol{\vartheta},f}$. We attract the reader's attention to the fact that, now, we add the index f to all f-dependent quantities; previously, this dependence was implicit as our models were fully parametric. As for the finite-dimensional setup discussed above, efficient inference procedures about $\boldsymbol{\vartheta}$ can be obtained by computing an efficient central sequence that takes care of the (infinite-dimensional) nuisance parameter f. In order to obtain such a central sequence, we here need to project $\boldsymbol{\Delta}^{(n)}_f(\boldsymbol{\vartheta})$ on the orthocomplement of the so-called *tangent space*. Such a tangent space can often be obtained by applying the following strategy. Let $\mathcal{Q}$ denote the set of all maps $q : (-1,1)^p \to \mathcal{F}$ such that $q(0) = f$. Then the tangent space can be obtained by computing the space generated by the gradient of $\log dP^{(n)}_{\boldsymbol{\vartheta},q(\boldsymbol{\xi})}$ with respect to $\boldsymbol{\xi}$ in the vicinity of $\boldsymbol{\xi} = 0$. Writing $\mathbf{L}^{(n)}_{q|f}(\boldsymbol{\vartheta})$ for the tangent space, an efficient central sequence can be defined as

$$\tilde{\boldsymbol{\Delta}}^{(n)}_{q|f}(\boldsymbol{\vartheta}) := Pr(\boldsymbol{\Delta}^{(n)}_f(\boldsymbol{\vartheta})|(\mathbf{L}^{(n)}_{q|f}(\boldsymbol{\vartheta}))^{\perp}].$$

The major problem of $\tilde{\boldsymbol{\Delta}}^{(n)}_{q|f}(\boldsymbol{\vartheta})$ is that it still depends on the choice of the surface (in a Hilbert space) indexed by $q \in \mathcal{Q}$. Therefore there is no guarantee that $\tilde{\boldsymbol{\Delta}}^{(n)}_{q|f}(\boldsymbol{\vartheta})$ will be orthogonal to $\mathbf{L}^{(n)}_{q'|f}(\boldsymbol{\vartheta})$ for every q' in $\mathcal{Q}$ and as a result semiparametric efficiency is not guaranteed either.

There are essentially two distinct approaches to deal with this particular problem. First, semiparametric efficiency can sometimes be reached by estimating the nuisance parameter f (Chen & Bickel 2006). Another possibility is to get rid of the nuisance f by invoking the *invariance principle*. We discuss this second solution in more detail. Consider the sequence of experiments

$$\mathcal{E}^{(n)} = \left(\mathcal{X}^{(n)}, \mathcal{A}^{(n)}, \mathcal{P}^{(n)} := \{P^{(n)}_{\boldsymbol{\vartheta},f} | f \in \mathcal{F}\}\right), \qquad (5.14)$$

associated with the fixed-$\boldsymbol{\vartheta}$ model and a sequence of σ-fields $\mathcal{B}^{(n)}(\boldsymbol{\vartheta}) \subset \mathcal{A}_n$ such that the restriction of $P^{(n)}_{\boldsymbol{\vartheta},f}$ to $\mathcal{B}^{(n)}(\boldsymbol{\vartheta})$, denoted as $P^{(n)}_{\boldsymbol{\vartheta},f|\mathcal{B}^{(n)}(\boldsymbol{\vartheta})}$, does not depend on $f \in \mathcal{F}$. The σ-fields $\mathcal{B}^{(n)}(\boldsymbol{\vartheta})$ are in general generated by the orbits of some group of transformations acting on $(\mathcal{X}^{(n)}, \mathcal{A}^{(n)})$. Hallin & Werker (2003) established that for any $q \in \mathcal{Q}$, $\mathbf{L}^{(n)}_{q|f}(\boldsymbol{\vartheta})$ is asymptotically orthogonal to $\mathcal{B}^{(n)}(\boldsymbol{\vartheta})$ in the sense that

$$\mathrm{E}_{\boldsymbol{\vartheta},f}[\mathbf{L}^{(n)}_{q|f}(\boldsymbol{\vartheta})|\mathcal{B}^{(n)}(\boldsymbol{\vartheta})] = o_{L^1}(1)$$

as $n \to \infty$ under $\mathrm{P}^{(n)}_{\boldsymbol{\vartheta},f}$ and therefore that asymptotic semiparametric efficiency is attained by considering the central sequence

$$\tilde{\boldsymbol{\Delta}}^{(n)}_{|f}(\boldsymbol{\vartheta}) := \mathrm{E}_{\boldsymbol{\vartheta},f}[\boldsymbol{\Delta}^{(n)}_{f}(\boldsymbol{\vartheta})|\mathcal{B}^{(n)}(\boldsymbol{\vartheta})].$$

To illustrate the use of the invariance principle, consider an n-tuple $Z_1, \ldots, Z_n$ of iid univariate random variables with an absolutely continuous (with respect to the Lebesgue measure) distribution, symmetric about some $\mu \in \mathbb{R}$. More precisely, assume that the common density of the Z_i's is of the form $f(x - \mu)$ where f belongs to the class $\mathcal{F}$ of positive densities that are symmetric around zero. We are interested in the problem of testing

$$\begin{cases} \mathcal{H}_0 : \mu = 0 \\ \mathcal{H}_1 : \mu \neq 0, \end{cases}$$

for which the infinite-dimensional parameter $f \in \mathcal{F}$ is a nuisance parameter. In order to eliminate the effect of this nuisance f, we consider the group of transformations $\mathcal{G} := \{G^{(n)}_{g_0}\}$ defined as

$$G^{(n)}_{g_0}(Z_1, \ldots, Z_n) = (g_0(Z_1), \ldots, g_0(Z_n)),$$

where g_0 is an odd, monotone increasing, continuous function such that $\lim_{z \to +\infty} g_0(z) = +\infty$ and $\lim_{z \to -\infty} g_0(z) = -\infty$. It is clear that if $Z_1, \ldots, Z_n$ is an n-tuple that belongs to $\mathcal{H}_0$, the n-tuple $g_0(Z_1), \ldots, g_0(Z_n)$ also belongs to $\mathcal{H}_0$. The transformations induced by $\mathcal{G}$ only act on the infinite-dimensional nuisance parameter f. The null hypothesis is therefore invariant with respect to $\mathcal{G}$. The invariance principle advocates test procedures that are invariant with respect to $\mathcal{G}$. Since every invariant statistic is measurable with respect to some maximal invariant[1] for $\mathcal{G}$, the resulting procedures have to be based on that maximal invariant. Letting $S_i = \mathrm{sign}(Z_i)$ and R_i be the rank of $|Z_i|$ among $|Z_1|, \ldots, |Z_n|$, it is easy to check that the maximal invariant associated with $\mathcal{G}$ is the vector

$$\mathbf{I}^{(n)} := (S_1, \ldots, S_n, R_1, \ldots, R_n)$$

[1]A function T is a maximal invariant for a transformation group $\mathcal{G}$ acting on $\mathcal{X}^{(n)}$ if $T(\mathbf{z}_1) = T(\mathbf{z}_2)$ iff there exists $g \in \mathcal{G}$ such that $\mathbf{z}_1 = g(\mathbf{z}_2)$ for all $\mathbf{z}_1, \mathbf{z}_2 \in \mathcal{X}^{(n)}$.

of signs and ranks. Since $\mathcal{G}$ is a generating group of the null hypothesis (i.e., it has a single orbit), procedures based on $\mathbf{I}^{(n)}$ will be distribution-free. Let $\mathcal{B}^{(n)}$ be the sigma-field associated with $\mathbf{I}^{(n)}$ and assume that the model is LAN with central sequence $\Delta_0^{(n)}$ at $\mu = 0$. Semiparametric efficiency can then be reached by considering test procedures based on $\mathrm{E}[\Delta_0^{(n)}|\mathcal{B}^{(n)}]$.

Example (continued). *In the case of a centered iid noise as in the example (5.2), consider the group of transformations* $\mathcal{G} := \{G_{g_0}^{(n)}\}$ *defined as*

$$G_{g_0}^{(n)}(Z_1^r(\boldsymbol{\beta}), \ldots, Z_n^r(\boldsymbol{\beta})) = (g_0(Z_1^r(\boldsymbol{\beta})), \ldots, g_0(Z_n^r(\boldsymbol{\beta}))),$$

where g_0 *is a monotone increasing, continuous function such that* $\lim_{z\to+\infty} g_0(z) = +\infty$, $\lim_{z\to-\infty} g_0(z) = -\infty$ *and* $g_0(0) = 0$. *The maximal invariant associated with this group of transformations is the vector of signs* $(S_1^{(n)}, \ldots, S_n^{(n)})$, *where* $S_i^{(n)} = \mathrm{sign}(Z_i^r(\boldsymbol{\beta}))$, *and ranks* $(R_1, \ldots, R_n)$, *where* R_i *is here the rank of* $Z_i^r(\boldsymbol{\beta})$ *among* $Z_1^r(\boldsymbol{\beta}), \ldots, Z_n^r(\boldsymbol{\beta})$. *The corresponding central sequence is therefore obtained as*

$$\mathrm{E}[\boldsymbol{\Delta}^{(n)}(\boldsymbol{\beta})|(S_1^{(n)}, \ldots, S_n^{(n)}, R_1^{(n)}, \ldots, R_n^{(n)})],$$

where the expectation is taken under $\mathrm{P}_{\boldsymbol{\beta},f}^{(n)}$.

5.3 LAN for directional data

The first extension of the Le Cam methodology to directional supports was provided by Ley et al. (2013), in the context of rotationally symmetric distributions (see Section 2.3.2 of Chapter 2). Proving that rotationally symmetric experiments are ULAN is far from trivial, the major difficulty being that the location parameter $\boldsymbol{\mu}$ belongs to the unit sphere $\mathcal{S}^{p-1}$ which is *curved.*

The solution of Ley et al. (2013) consists in (i) expressing all quantities in spherical coordinates which are defined on a linear domain, (ii) establishing the ULAN property for these new parameters, and (iii) translating the ULAN property into the original parameterization by using a lemma proved in Hallin et al. (2010). With this "directional ULAN property" in hand, Ley et al. (2013) constructed optimal R-estimators for the location parameter of rotationally symmetric distributions. A more general version of this property, in the case of m independent populations, permitted Ley et al. (2017) to propose efficient ANOVA for spherical distributions.

In what follows, we present some of the results obtained in Ley et al. (2013) and Ley et al. (2017). We also illustrate the usefulness of the Third Le Cam Lemma by computing the local powers of tests for the concentration parameter of Fisher–von Mises–Langevin distributions (Ley & Verdebout 2014*a*). However, we start by summarizing the general results related to curved experiments obtained in Hallin et al. (2010).

5.3.1 The Le Cam methodology for curved experiments and associated efficient tests

We discuss here a general theory developed in Hallin et al. (2010) for locally asymptotically optimal tests in curved ULAN experiments. Consider a sequence of experiments

$$\mathcal{E}_1^{(n)} = \left(\mathcal{X}^{(n)}, \mathcal{A}^{(n)}, \mathcal{P}^{(n)} := \{\mathrm{P}_{\boldsymbol{\omega}}^{(n)} : \boldsymbol{\omega} \in \boldsymbol{\Omega}\}\right),$$

where $\boldsymbol{\Omega}$ is an open subset of $\mathbb{R}^{p_1}$. Let

$$\mathcal{E}_2^{(n)} := \left(\mathcal{X}^{(n)}, \mathcal{A}^{(n)}, \mathcal{P}^{(n)} := \{\mathrm{P}_{\boldsymbol{\vartheta}}^{(n)} : \boldsymbol{\vartheta} \in \boldsymbol{\Theta} := \vec{d}(\boldsymbol{\Omega})\}\right),$$

where the mapping $\vec{d} : \mathbb{R}^{p_1} \to \mathbb{R}^{p_2}$ $(p_1 \leq p_2)$ is of full column rank so that in general $\boldsymbol{\Theta}$ is possibly a non-linear manifold of $\mathbb{R}^{p_2}$. In such a case, the experiments $\mathcal{E}_2^{(n)}$ are said to be curved. We observe that the experiments $\mathcal{E}_1^{(n)}$ and $\mathcal{E}_2^{(n)}$ are virtually the same but are indexed by different parameterizations through $\boldsymbol{\vartheta} = \vec{d}(\boldsymbol{\omega})$. If the sequence of experiments $\mathcal{E}_1^{(n)}$ is ULAN, then $\mathcal{E}_2^{(n)}$ is also ULAN as shown in the following result.

Lemma 5.3.1 (Hallin et al. 2010) *Consider a family of probability distributions* $\mathcal{P}^{(n)} = \{\mathrm{P}_{\boldsymbol{\omega}}^{(n)} \mid \boldsymbol{\omega} \in \boldsymbol{\Omega}\}$ *with* $\boldsymbol{\Omega}$ *an open subset of* $\mathbb{R}^{p_1}$ $(p_1 \in \mathbb{N}_0)$. *Suppose that the parameterization* $\boldsymbol{\omega} \mapsto \mathrm{P}_{\boldsymbol{\omega}}^{(n)}$ *is ULAN for* $\mathcal{P}^{(n)}$ *at some point* $\boldsymbol{\omega}_0 \in \Omega$, *with central sequence* $\boldsymbol{\Delta}^{(n)}(\boldsymbol{\omega}_0)$ *and Fisher information matrix* $\boldsymbol{\Gamma}(\boldsymbol{\omega}_0)$. *Let* $\vec{d} : \boldsymbol{\omega} \mapsto \boldsymbol{\vartheta} := \vec{d}(\boldsymbol{\omega})$ *be a continuously differentiable mapping from* $\mathbb{R}^{p_1}$ *to* $\mathbb{R}^{p_2}$ $(p_1 \leq p_2 \in \mathbb{N}_0)$ *with full column rank Jacobian matrix* $D\vec{d}(\boldsymbol{\omega})$ *at every* $\boldsymbol{\omega}$ *in some neighborhood of* $\boldsymbol{\omega}_0$. *Write* $\boldsymbol{\Theta} := \vec{d}(\boldsymbol{\Omega})$, *and assume that* $\boldsymbol{\vartheta} \mapsto \mathrm{P}_{\boldsymbol{\vartheta}}^{(n);\vec{d}}$, $\boldsymbol{\vartheta} \in \boldsymbol{\Theta}$, *provides another parameterization of* $\mathcal{P}^{(n)}$. *Then* $\boldsymbol{\vartheta} \mapsto \mathrm{P}_{\boldsymbol{\vartheta}}^{(n);\vec{d}}$ *is also ULAN for* $\mathcal{P}^{(n)}$ *at* $\boldsymbol{\vartheta}_0 = \vec{d}(\boldsymbol{\omega}_0)$, *with central sequence* $\boldsymbol{\Delta}^{(n);\vec{d}}(\boldsymbol{\vartheta}_0) = (D^{-}\vec{d}(\boldsymbol{\omega}_0))'\boldsymbol{\Delta}^{(n)}(\boldsymbol{\omega}_0)$ *and Fisher information matrix* $\boldsymbol{\Gamma}^{\vec{d}}(\boldsymbol{\vartheta}_0) = (D^{-}\vec{d}(\boldsymbol{\omega}_0))'\boldsymbol{\Gamma}(\boldsymbol{\omega}_0)D^{-}\vec{d}(\boldsymbol{\omega}_0)$, *where* $D^{-}\vec{d}(\boldsymbol{\omega}_0) := ((D\vec{d}(\boldsymbol{\omega}_0))'D\vec{d}(\boldsymbol{\omega}_0))^{-1}(D\vec{d}(\boldsymbol{\omega}_0))'$ *is the Moore–Penrose inverse of* $D\vec{d}(\boldsymbol{\omega}_0)$.

Now, let us consider a sequence of curved ULAN experiments $\mathcal{E}_2^{(n)} = \left(\mathcal{X}^{(n)}, \mathcal{A}^{(n)}, \mathcal{P}^{(n)} := \{\mathrm{P}_{\boldsymbol{\vartheta}}^{(n)} : \boldsymbol{\vartheta} \in \boldsymbol{\Theta}\}\right)$ with $\boldsymbol{\Theta}$ a non-linear manifold of $\mathbb{R}^p$. Besides Lemma 5.3.1, Hallin et al. (2010) provide a general method to construct locally and asymptotically most stringent tests for null hypotheses of the form

$$\mathcal{H}_0 : \boldsymbol{\vartheta} \in C \cap \boldsymbol{\Theta},$$

where C is an r-dimensional manifold in $\mathbb{R}^p$, $r < p$. To obtain a locally (at $\boldsymbol{\vartheta}_0 \in \mathcal{H}_0$) and asymptotically optimal test in the $\boldsymbol{\vartheta}$-parameterization, one has to find a local chart $\mathfrak{b} : A \subseteq \mathbb{R}^r \to \mathbb{R}^p$ (at $\boldsymbol{\vartheta}_0$) for the manifold $C \cap \boldsymbol{\Theta}$ and establish the ULAN property in the simpler parameterization $\boldsymbol{\eta}_0 := \mathfrak{b}^{-1}(\boldsymbol{\vartheta}_0)$. Then, letting $\boldsymbol{\Delta}^{(n)}(\boldsymbol{\vartheta}_0)$ and $\boldsymbol{\Gamma}(\boldsymbol{\vartheta}_0)$ be, respectively, the central sequence and the Fisher information associated with the $\boldsymbol{\Theta}$-experiments $\mathcal{E}_2^{(n)}$, the test that rejects the null when ($\mathbf{A}^-$ stands for the Moore–Penrose pseudo-inverse of $\mathbf{A}$ and $D\mathfrak{b}(\boldsymbol{\eta}_0)$ is the Jacobian matrix of $\mathfrak{b}$ computed at $\boldsymbol{\eta}_0$)

$$Q_{\boldsymbol{\vartheta}_0}^{(n)} := (\boldsymbol{\Delta}_{\boldsymbol{\vartheta}_0}^{(n)})'(\boldsymbol{\Gamma}_{\boldsymbol{\vartheta}_0}^{-} - D\mathfrak{b}(\boldsymbol{\eta}_0)\,(D\mathfrak{b}'(\boldsymbol{\eta}_0)\boldsymbol{\Gamma}_{\boldsymbol{\vartheta}_0} D\mathfrak{b}(\boldsymbol{\eta}_0))^{-} D\mathfrak{b}'(\boldsymbol{\eta}_0))\boldsymbol{\Delta}_{\boldsymbol{\vartheta}_0}^{(n)} \qquad (5.15)$$

exceeds the α-upper quantile of the chi-squared distribution with $p - r$ degrees of freedom is locally and asymptotically most stringent for $\mathcal{H}_0 : \boldsymbol{\vartheta} \in C \cap \boldsymbol{\Theta}$ against $\mathcal{H}_1 : \boldsymbol{\vartheta} \notin C \cap \boldsymbol{\Theta}$.

5.3.2 LAN property for rotationally symmetric distributions

Throughout this section, let $\mathbf{X}_1, \ldots, \mathbf{X}_n$ be a sequence of iid rotationally symmetric observations. We mostly consider here statistical inference for the location parameter $\boldsymbol{\mu} \in \mathcal{S}^{p-1}$ of a rotationally symmetric distribution. Estimation and testing procedures for $\boldsymbol{\mu}$ have been extensively studied in the literature, with much of the focus in the past years being put on the class of M-estimators. An M-estimator $\hat{\boldsymbol{\mu}}$ associated with a given function $\rho_0(\mathbf{x}; \boldsymbol{\mu})$ is defined as the value of $\boldsymbol{\mu}$ that minimizes the objective function

$$\boldsymbol{\mu} \mapsto \rho(\boldsymbol{\mu}) := \sum_{i=1}^{n} \rho_0(\mathbf{X}_i; \boldsymbol{\mu}).$$

These M-estimators are robust to outliers and enjoy nice asymptotic properties, see Chang & Rivest (2001), Chang & Tsai (2003), or Chang (2004). In particular, the choice $\rho_0(\mathbf{x}; \boldsymbol{\mu}) = \arccos(\mathbf{x}'\boldsymbol{\mu})$ yields the spherical median of Fisher (1985), whereas $\rho_0(\mathbf{x}; \boldsymbol{\mu}) = \|\mathbf{x} - \boldsymbol{\mu}\|^2$ yields $\hat{\boldsymbol{\mu}} = \bar{\mathbf{X}}/\|\bar{\mathbf{X}}\|$, the spherical mean.

Recently Ley et al. (2013) and Paindaveine & Verdebout (2015) considered locally asymptotically optimal inference for $\boldsymbol{\mu}$. The techniques used in both papers are based on the general results presented in the previous section. The various results provided in what follows are based on the following assumption.

ASSUMPTION A. $\mathbf{X}_1, \ldots, \mathbf{X}_n$ are iid with common distribution $\mathrm{P}_{\boldsymbol{\mu},f}$ characterized by a density of the form

$$\mathbf{x} \mapsto f_{\boldsymbol{\mu}}(\mathbf{x}) = c_{p,f}\, f(\mathbf{x}'\boldsymbol{\mu}), \quad \mathbf{x} \in \mathcal{S}^{p-1}, \tag{5.16}$$

where $\boldsymbol{\mu} \in \mathcal{S}^{p-1}$ is the location parameter and the angular function $f : [-1,1] \to \mathbb{R}_0^+$ is absolutely continuous and monotone non-decreasing.

We denote here by $\mathcal{F}$ the set of functions f satisfying Assumption A. Section 2.3.2 of Chapter 2 gives numerous examples of angular functions f and provides further insight into rotationally symmetric distributions. For the sake of readability, we restate here some important results, but refer to Section 2.3.2 for the details. If $\mathbf{X}_1, \ldots, \mathbf{X}_n$ are iid with density (5.16), then $\mathbf{X}_1'\boldsymbol{\mu}, \ldots, \mathbf{X}_n'\boldsymbol{\mu}$ are iid with density

$$t \mapsto \tilde{f}(t) := \tilde{c}_{p,f} f(t)(1-t^2)^{(p-3)/2}, \quad -1 \leq t \leq 1, \tag{5.17}$$

where $\tilde{c}_{p,f}$ is a normalizing constant. Rotational symmetry implies that, for each $i = 1, \ldots, n$, $\mathbf{X}_i'\boldsymbol{\mu}$ and the multivariate sign

$$\mathbf{S}_{\boldsymbol{\mu}}(\mathbf{X}_i) = \frac{\mathbf{X}_i - (\mathbf{X}_i'\boldsymbol{\mu})\boldsymbol{\mu}}{\|\mathbf{X}_i - (\mathbf{X}_i'\boldsymbol{\mu})\boldsymbol{\mu}\|}$$

are independent. Furthermore, $\mathbf{S}_{\boldsymbol{\mu}}(\mathbf{X}_i)$ is uniformly distributed on the sphere $\mathcal{S}^{p-1}(\boldsymbol{\mu}^{\perp}) := \{\mathbf{v} \in \mathbb{R}^p \,|\, \mathbf{v}'\mathbf{v} = 1, \mathbf{v}'\boldsymbol{\mu} = 0\}$. In the sequel we let $\mathrm{P}^{(n)}_{\boldsymbol{\mu},f}$ denote the joint distribution of $\mathbf{X}_1, \ldots, \mathbf{X}_n$ under Assumption A.

In Ley et al. (2013), it is shown that the rotationally symmetric model is ULAN (with respect to $\boldsymbol{\mu}$). This requires another technical assumption.

ASSUMPTION B. Letting $\varphi_f := \dot{f}/f$ ($\dot{f}$ is the a.e.-derivative of f), the quantity $\mathcal{J}_p(f) := \int_{-1}^{1} \varphi_f^2(t)(1-t^2)\tilde{f}(t)dt < +\infty$.

Let $\boldsymbol{\mu}^{(n)} \in \mathcal{S}^{p-1}$ be such that $\boldsymbol{\mu}^{(n)} - \boldsymbol{\mu} = O(n^{-1/2})$ and consider local alternatives on the sphere of the form $\boldsymbol{\mu}^{(n)} + n^{-1/2}\mathbf{t}^{(n)}$ with $\mathbf{t}^{(n)} \in \mathbb{R}^p$ a bounded sequence. For $\boldsymbol{\mu}^{(n)} + n^{-1/2}\mathbf{t}^{(n)}$ to remain in $\mathcal{S}^{p-1}$, it is necessary that the sequence $\mathbf{t}^{(n)}$ satisfies

$$\begin{aligned} 0 &= (\boldsymbol{\mu}^{(n)} + n^{-1/2}\mathbf{t}^{(n)})'(\boldsymbol{\mu}^{(n)} + n^{-1/2}\mathbf{t}^{(n)}) - 1 \\ &= 2n^{-1/2}(\boldsymbol{\mu}^{(n)})'\mathbf{t}^{(n)} + n^{-1}(\mathbf{t}^{(n)})'\mathbf{t}^{(n)}. \end{aligned} \tag{5.18}$$

Consequently, $\mathbf{t}^{(n)}$ must be such that $2n^{-1/2}(\boldsymbol{\mu}^{(n)})'\mathbf{t}^{(n)} + o(n^{-1/2}) = 0$. Therefore, for $\boldsymbol{\mu}^{(n)} + n^{-1/2}\mathbf{t}^{(n)}$ to remain in $\mathcal{S}^{p-1}$, $\mathbf{t}^{(n)}$ must belong, up to a $o(n^{-1/2})$ quantity, to the tangent space to $\mathcal{S}^{p-1}$ at $\boldsymbol{\mu}^{(n)}$. The following result can then be established.

Proposition 5.3.1 (Ley et al. 2013) *Let Assumptions A and B hold. Then the family of probability distributions* $\left\{\mathrm{P}^{(n)}_{\boldsymbol{\mu},f} \mid \boldsymbol{\mu} \in \mathcal{S}^{p-1}\right\}$ *is ULAN with central sequence*

$$\boldsymbol{\Delta}^{(n)}_{\boldsymbol{\mu};f} := n^{-1/2}\sum_{i=1}^{n} \varphi_f(\mathbf{X}_i'\boldsymbol{\mu})(1-(\mathbf{X}_i'\boldsymbol{\mu})^2)^{1/2}\mathbf{S}_{\boldsymbol{\mu}}(\mathbf{X}_i)$$

and Fisher information matrix

$$\boldsymbol{\Gamma}_{\boldsymbol{\mu};f} := \frac{\mathcal{J}_p(f)}{p-1}(\mathbf{I}_p - \boldsymbol{\mu}\boldsymbol{\mu}').$$

More precisely, for any $\boldsymbol{\mu}^{(n)} \in \mathcal{S}^{p-1}$ *such that* $\boldsymbol{\mu}^{(n)} - \boldsymbol{\mu} = O(n^{-1/2})$ *and any bounded sequence* $\mathbf{t}^{(n)}$ *as in (5.18), we have*

$$\log\left(\frac{d\mathrm{P}^{(n)}_{\boldsymbol{\mu}^{(n)}+n^{-1/2}\mathbf{t}^{(n)},f}}{d\mathrm{P}^{(n)}_{\boldsymbol{\mu}^{(n)},f}}\right) = (\mathbf{t}^{(n)})'\boldsymbol{\Delta}^{(n)}_{\boldsymbol{\mu}^{(n)};f} - \frac{1}{2}(\mathbf{t}^{(n)})'\boldsymbol{\Gamma}_{\boldsymbol{\mu};f}\mathbf{t}^{(n)} + o_{\mathrm{P}}(1)$$

and $\boldsymbol{\Delta}^{(n)}_{\boldsymbol{\mu}^{(n)};f} \xrightarrow{\mathcal{D}} \mathcal{N}_p(\mathbf{0}, \boldsymbol{\Gamma}_{\boldsymbol{\mu};f})$, *both under* $\mathrm{P}^{(n)}_{\boldsymbol{\mu}^{(n)},f}$, *as* $n \to \infty$.

This result was used in Ley et al. (2013), Paindaveine & Verdebout (2015) and Ley et al. (2017) to obtain locally and asymptotically optimal estimators and tests based on signed-ranks that combine semiparametric efficiency and invariance as described in Section 5.2.4. We briefly discuss these results in the next two sections.

5.3.3 Application 1: Optimal inference based on signed-ranks

Fix $\boldsymbol{\mu} \in \mathcal{S}^{p-1}$ and consider the family $\bigcup_{f\in\mathcal{F}} \mathrm{P}^{(n)}_{\boldsymbol{\mu},f}$. We recall the tangent normal decomposition (2.21):

$$\mathbf{X}_i = (\mathbf{X}_i'\boldsymbol{\mu})\boldsymbol{\mu} + \sqrt{1-(\mathbf{X}_i'\boldsymbol{\mu})^2}\mathbf{S}_{\boldsymbol{\mu}}(\mathbf{X}_i).$$

Now, let $\mathcal{G}_h^{(n)}$ be the group of transformations of the form $g_h^{(n)} : (\mathbf{X}_1, \ldots, \mathbf{X}_n) \mapsto (g_h(\mathbf{X}_1), \ldots, g_h(\mathbf{X}_n))$ with

$$g_h(\mathbf{X}_i) := h(\mathbf{X}_i'\boldsymbol{\mu})\boldsymbol{\mu} + \sqrt{1-h(\mathbf{X}_i'\boldsymbol{\mu})^2}\mathbf{S}_{\boldsymbol{\mu}}(\mathbf{X}_i), \quad i = 1, \ldots, n, \qquad (5.19)$$

where $h : [-1,1] \to [-1,1]$ is a continuous monotone non-decreasing function such that $h(1) = 1$ and $h(-1) = -1$. For any $g_h^{(n)} \in \mathcal{G}_h^{(n)}$, it is easy to verify that $\|g_h^{(n)}(\mathbf{X}_i)\| = 1$, meaning that $g_h^{(n)} \in \mathcal{G}_h^{(n)}$ is a monotone transformation from $(\mathcal{S}^{p-1})^n$ to $(\mathcal{S}^{p-1})^n$. The group $\mathcal{G}_h^{(n)}$ is a generating group of the family of distributions $\bigcup_{f\in\mathcal{F}} \mathrm{P}_{\boldsymbol{\mu},f}^{(n)}$.

These considerations naturally raise the issue of finding the maximal invariant associated with $\mathcal{G}_h^{(n)}$. Note that it is easy to verify that $\mathbf{S}_{\boldsymbol{\mu}}(g_h(\mathbf{X}_i)) = \mathbf{S}_{\boldsymbol{\mu}}(\mathbf{X}_i)$ so that the vector of signs $\mathbf{S}_{\boldsymbol{\mu}}(\mathbf{X}_1), \ldots, \mathbf{S}_{\boldsymbol{\mu}}(\mathbf{X}_n)$ is invariant under the action of $\mathcal{G}_h^{(n)}$. However, it is not a maximal invariant; indeed, the latter is composed of signs *and* ranks. Define, for all $i = 1, \ldots, n$, R_i as the rank of $\mathbf{X}_i'\boldsymbol{\mu}$ among $\mathbf{X}_1'\boldsymbol{\mu}, \ldots, \mathbf{X}_n'\boldsymbol{\mu}$. The fact that $g_h(\mathbf{X}_i)'\boldsymbol{\mu} = h(\mathbf{X}_i'\boldsymbol{\mu})$ directly implies the invariance of these new ranks under the action of the group $\mathcal{G}_h^{(n)}$. The maximal invariant associated with $\mathcal{G}_h^{(n)}$ thus corresponds to the combination of the signs $\mathbf{S}_{\boldsymbol{\mu}}(\mathbf{X}_1), \ldots, \mathbf{S}_{\boldsymbol{\mu}}(\mathbf{X}_n)$ with the ranks $R_1, \ldots, R_n$. It follows that any statistic measurable with respect to the signs $\mathbf{S}_{\boldsymbol{\mu}}(\mathbf{X}_i)$ and ranks R_i is distribution-free under $\bigcup_{f\in\mathcal{F}} \mathrm{P}_{\boldsymbol{\mu},f}^{(n)}$.

In accordance with these findings, R-estimation in Ley et al. (2013) and signed-rank tests in Paindaveine & Verdebout (2015) are based on a signed-rank version of the parametric central sequence obtained in Proposition 5.3.1, namely on

$$\underset{\sim}{\boldsymbol{\Delta}}_{\boldsymbol{\mu};K}^{(n)} := n^{-1/2} \sum_{i=1}^{n} K\left(\frac{R_i}{n+1}\right) \mathbf{S}_{\boldsymbol{\mu}}(\mathbf{X}_i), \tag{5.20}$$

where K is a score function satisfying some regularity conditions. Letting $\hat{\boldsymbol{\mu}}^{(n)}$ stand for a preliminary root-n consistent and locally discrete estimator,[2] the proposed R-estimators are of the form

$$\hat{\boldsymbol{\mu}}_K^{(n)} := \frac{\hat{\boldsymbol{\mu}}^{(n)} + (p-1)(\hat{\mathcal{J}}(K,g))^{-1} \underset{\sim}{\boldsymbol{\Delta}}_{\hat{\boldsymbol{\mu}}^{(n)};K}^{(n)}}{\|\hat{\boldsymbol{\mu}}^{(n)} + (p-1)(\hat{\mathcal{J}}(K,g))^{-1} \underset{\sim}{\boldsymbol{\Delta}}_{\hat{\boldsymbol{\mu}}^{(n)};K}^{(n)}\|},$$

where g is an angular function as in Assumption A and $\hat{\mathcal{J}}(K,g)$ is a consistent estimator of the cross-information quantity $\mathcal{J}(K,g) := \int_0^1 K(u)K_g(u)du$ with $K_g(u) := \varphi_g(\tilde{G}^{-1}(u))(1-(\tilde{G}^{-1}(u))^2)^{1/2}$. Here $\tilde{G}$ is the cumulative distribution function associated with $\tilde{g}$ given by (5.17). When based on K_f, $\hat{\boldsymbol{\mu}}_{K_f}$ is asymptotically efficient under $\mathrm{P}_{\boldsymbol{\mu},f}^{(n)}$. These estimators are called one-step estimators, because

[2]This means that $\hat{\boldsymbol{\mu}}^{(n)}$ only takes a bounded number of distinct values in $\boldsymbol{\mu}$-centered balls with $O\left(n^{-1/2}\right)$ radius. This discretization condition is a purely technical requirement with little practical implications.

they update any preliminary root-n consistent estimator and turn it into an efficient one.

In Paindaveine & Verdebout (2015), the authors provided signed-rank based tests that reject $\mathcal{H}_0 : \boldsymbol{\mu} = \boldsymbol{\mu}_0$ at asymptotic level α when

$$T_K := \frac{(p-1)}{\mathcal{J}(K)} \|\underset{\sim}{\boldsymbol{\Delta}}^{(n)}_{\boldsymbol{\mu}_0;K}\|^2 > \chi^2_{p-1;1-\alpha}$$

where $\mathcal{J}(K) := \int_0^1 K^2(u)du$. When based on K_f, T_{K_f} is locally and asymptotically optimal under $\mathrm{P}^{(n)}_{\boldsymbol{\mu},f}$.

5.3.4 Application 2: ANOVA on spheres

Let us now consider $m(\geq 2)$ mutually independent samples $\mathbf{X}_{i1}, \ldots, \mathbf{X}_{in_i}$, $i = 1, \ldots, m$, of rotationally symmetric observations on $\mathcal{S}^{p-1}$ satisfying

ASSUMPTION A. For all $i = 1, \ldots, m$, $\mathbf{X}_{i1}, \ldots, \mathbf{X}_{in_i}$ are iid with joint distribution $\mathrm{P}^{(n)}_{\boldsymbol{\mu}_i;f_i}$, with location parameter $\boldsymbol{\mu}_i \in \mathcal{S}^{p-1}$ and angular function $f_i : [-1, 1] \to \mathbb{R}^+_0$ an absolutely continuous and monotone non-decreasing mapping.

ASSUMPTION B. The Fisher information associated with the location parameter $\boldsymbol{\mu}$ is finite; this finiteness is ensured if, for $i = 1, \ldots, m$ and letting $\varphi_{f_i} := \dot{f}_i/f_i$ ($\dot{f}_i$ is the a.e.-derivative of f_i), $\mathcal{J}_p(f_i) := \int_{-1}^1 \varphi^2_{f_i}(t)(1-t^2)\tilde{f}_i(t)dt < +\infty$.

In order to be able to state the ULAN result below, we need to impose some conditions on the sample sizes n_i, $i = 1, \ldots, m$. This we achieve via the following

ASSUMPTION C. Letting $n = \sum_{i=1}^m n_i$, for all $i = 1, \ldots, m$ the ratio $r_i^{(n)} := n_i/n$ converges to a finite constant r_i as $n \to \infty$. A matrix made up of these constants is defined as

$$\mathbf{r}^{(n)} := \mathrm{diag}\left(\left(r_1^{(n)}\right)^{-1/2} \mathbf{I}_p, \ldots, \left(r_m^{(n)}\right)^{-1/2} \mathbf{I}_p\right).$$

Throughout this section, we denote by $\mathcal{F}^m$ the collection of m-tuples of angular functions $\underline{f} := (f_1, f_2, \ldots, f_m)$, by $\boldsymbol{\vartheta} := (\boldsymbol{\mu}_1', \ldots, \boldsymbol{\mu}_m')'$ the parameter of interest and, consequently, by $\mathrm{P}^{(n)}_{\boldsymbol{\vartheta},\underline{f}}$ the joint distribution of all n observations. Under Assumptions A, B and C, the objective of Ley et al. (2017) was to address the spherical ANOVA problem $\mathcal{H}_0 : \boldsymbol{\mu}_1 = \ldots = \boldsymbol{\mu}_m$ versus the alternative $\mathcal{H}_1 : \exists 1 \leq i \neq j \leq m$ for which $\boldsymbol{\mu}_i \neq \boldsymbol{\mu}_j$. This is a problem of capital importance in paleomagnetism, see Section 1.2.1 of the Introduction. The proposed tests are based on the following extension of the ULAN property from Proposition 5.3.1.

Proposition 5.3.2 (Ley et al. 2017) *Let Assumptions A, B and C hold. Then the model* $\left\{\mathrm{P}^{(n)}_{\boldsymbol{\vartheta},\underline{f}} \mid \boldsymbol{\vartheta} \in (\mathcal{S}^{p-1})^m\right\}$ *is ULAN with central sequence* $\boldsymbol{\Delta}^{(n)}_{\boldsymbol{\vartheta};\underline{f}} := \left((\boldsymbol{\Delta}^{(n)}_{\boldsymbol{\mu}_1;f_1})', \ldots, (\boldsymbol{\Delta}^{(n)}_{\boldsymbol{\mu}_m;f_m})'\right)'$*, where*

$$\boldsymbol{\Delta}^{(n)}_{\boldsymbol{\mu}_i;f_i} := n_i^{-1/2} \sum_{j=1}^{n_i} \varphi_{f_i}(\mathbf{X}'_{ij}\boldsymbol{\mu}_i)(1-(\mathbf{X}'_{ij}\boldsymbol{\mu}_i)^2)^{1/2}\mathbf{S}_{\boldsymbol{\mu}_i}(\mathbf{X}_{ij}), \quad i = 1, \ldots, m,$$

and Fisher information matrix $\boldsymbol{\Gamma}_{\boldsymbol{\vartheta};\underline{f}} := \mathrm{diag}(\boldsymbol{\Gamma}_{\boldsymbol{\mu}_1;f_1}, \ldots, \boldsymbol{\Gamma}_{\boldsymbol{\mu}_m;f_m})$ *where*

$$\boldsymbol{\Gamma}_{\boldsymbol{\mu}_i;f_i} := \frac{\mathcal{J}_p(f_i)}{p-1}(\mathbf{I}_p - \boldsymbol{\mu}_i\boldsymbol{\mu}_i'), \quad i = 1, \ldots, m.$$

More precisely, for any $\boldsymbol{\vartheta}^{(n)} \in (\mathcal{S}^{p-1})^m$ *such that* $\boldsymbol{\vartheta}^{(n)} - \boldsymbol{\vartheta} = O(n^{-1/2})$ *and any bounded sequence* $\mathbf{t}^{(n)} = (\mathbf{t}_1^{(n)\prime}, \ldots, \mathbf{t}_m^{(n)\prime})' \in \mathbb{R}^{pm}$ *as in (5.18), we have*

$$\log\left(\frac{d\mathrm{P}^{(n)}_{\boldsymbol{\vartheta}^{(n)}+n^{-1/2}\mathbf{r}^{(n)}\mathbf{t}^{(n)},\underline{f}}}{d\mathrm{P}^{(n)}_{\boldsymbol{\vartheta}^{(n)},\underline{f}}}\right) = (\mathbf{t}^{(n)})'\boldsymbol{\Delta}^{(n)}_{\boldsymbol{\vartheta}^{(n)};\underline{f}} - \frac{1}{2}(\mathbf{t}^{(n)})'\boldsymbol{\Gamma}_{\boldsymbol{\vartheta};\underline{f}}\mathbf{t}^{(n)} + o_{\mathrm{P}}(1) \tag{5.21}$$

and $\boldsymbol{\Delta}^{(n)}_{\boldsymbol{\vartheta}^{(n)};\underline{f}} \xrightarrow{\mathcal{D}} \mathcal{N}_{mp}(\mathbf{0}, \boldsymbol{\Gamma}_{\boldsymbol{\vartheta};\underline{f}})$*, both under* $\mathrm{P}^{(n)}_{\boldsymbol{\vartheta},\underline{f}}$*, as* $n \to \infty$*.*

Proposition 5.3.2 provides all the necessary tools for building optimal procedures for the ANOVA testing problem. The related null hypothesis is the intersection between $(\mathcal{S}^{p-1})^m$ and the linear subspace (of $\mathbb{R}^{mp}$)

$$\mathcal{C} := \{\mathbf{v} = (\mathbf{v}_1', \ldots, \mathbf{v}_m')' \mid \mathbf{v}_1, \ldots, \mathbf{v}_m \in \mathbb{R}^p \text{ and } \mathbf{v}_1 = \ldots = \mathbf{v}_m\} =: \mathcal{M}(1_m \otimes \mathbf{I}_p)$$

where we set $1_m := (1, \ldots, 1)' \in \mathbb{R}^m$ and $\mathbf{A} \otimes \mathbf{B}$ denotes the Kronecker product between $\mathbf{A}$ and $\mathbf{B}$. Such a restriction, namely an intersection between a linear subspace and a non-linear manifold, is exactly the type of testing problems discussed in Section 5.3.1. As explained in that section, one has to consider the locally and asymptotically most stringent test for the null hypothesis defined by the intersection between $\mathcal{C}$ and the tangent to $(\mathcal{S}^{p-1})^m$. Let $\boldsymbol{\mu}$ denote the common value of $\boldsymbol{\mu}_1, \ldots, \boldsymbol{\mu}_m$ under the null hypothesis. In the vicinity of $1_m \otimes \boldsymbol{\mu}$, the intersection between $\mathcal{C}$ and the tangent to $(\mathcal{S}^{p-1})^m$ is given by

$$\Big\{(\boldsymbol{\mu}' + n^{-1/2}(r_1^{(n)})^{-1/2}\mathbf{t}_1^{(n)\prime}, \ldots, \boldsymbol{\mu}' + n^{-1/2}(r_m^{(n)})^{-1/2}\mathbf{t}_m^{(n)\prime})', \tag{5.22}$$
$$\boldsymbol{\mu}'\mathbf{t}_1^{(n)} = \ldots = \boldsymbol{\mu}'\mathbf{t}_m^{(n)} = 0, (r_1^{(n)})^{-1/2}\mathbf{t}_1^{(n)} = \ldots = (r_m^{(n)})^{-1/2}\mathbf{t}_m^{(n)}\Big\}.$$

Solving the system (5.22) yields

$$\mathbf{r}^{(n)}\mathbf{t}^{(n)} = \left((r_1^{(n)})^{-1/2}\mathbf{t}_1^{(n)\prime}, \ldots, (r_m^{(n)})^{-1/2}\mathbf{t}_m^{(n)\prime}\right)' \in \mathcal{M}(1_m \otimes (\mathbf{I}_p - \boldsymbol{\mu}\boldsymbol{\mu}')). \tag{5.23}$$

Loosely speaking, we have converted the initial null hypothesis $\mathcal{H}_0$ into a linear restriction of the form (5.23) in terms of local perturbations $\mathbf{t}^{(n)}$, for which Le Cam's asymptotic theory then provides a locally and asymptotically optimal parametric test under fixed $\underline{f}$. Using Proposition 5.3.2 and letting $\boldsymbol{\Upsilon}_{\boldsymbol{\vartheta}} := 1_m \otimes (\mathbf{I}_p - \boldsymbol{\mu}\boldsymbol{\mu}')$ and $\boldsymbol{\Upsilon}_{\boldsymbol{\vartheta};\mathbf{r}}^{(n)} := (\mathbf{r}^{(n)})^{-1}\boldsymbol{\Upsilon}_{\boldsymbol{\vartheta}}$, an asymptotically most stringent test $\phi_{\underline{f}}^{(n)}$ is then obtained by rejecting $\mathcal{H}_0$ at asymptotic level α if

$$Q_{\underline{f}}^{(n)} := (\boldsymbol{\Delta}_{\boldsymbol{\vartheta};\underline{f}}^{(n)})' \left(\boldsymbol{\Gamma}_{\boldsymbol{\vartheta};\underline{f}}^{-} - \boldsymbol{\Upsilon}_{\boldsymbol{\vartheta};\mathbf{r}}^{(n)} \left((\boldsymbol{\Upsilon}_{\boldsymbol{\vartheta};\mathbf{r}}^{(n)})'\boldsymbol{\Gamma}_{\boldsymbol{\vartheta};\underline{f}}\boldsymbol{\Upsilon}_{\boldsymbol{\vartheta};\mathbf{r}}^{(n)}\right)^{-} (\boldsymbol{\Upsilon}_{\boldsymbol{\vartheta};\mathbf{r}}^{(n)})'\right) \boldsymbol{\Delta}_{\boldsymbol{\vartheta};\underline{f}}^{(n)} > \chi^2_{(m-1)(p-1);1-\alpha}. \tag{5.24}$$

The parametric test $\phi_{\underline{f}}^{(n)}$ is locally and asymptotically optimal but nevertheless suffers from some drawbacks. First it is only valid under the prespecified m-tuple $\underline{f} = (f_1, \ldots, f_m)$. Since it is highly unrealistic in practice to assume that the underlying densities are known, these tests are useless for practitioners. Second, it assumes the value of the common location parameter $\boldsymbol{\mu}$ to be known, which is unrealistic, too. In practice, $\boldsymbol{\mu}$ has to be estimated. Both issues are taken care of by Ley et al. (2017), who proposed *pseudo-FvML* tests and rank-based versions of $\phi_{\underline{f}}^{(n)}$.

The idea underpinning the pseudo-FvML test is similar in flavor to the *pseudo-Gaussian* tests developed in the classical "linear" framework in Hallin & Paindaveine (2008). More specifically, the approach employed makes use of the FvML as basis distribution, building the (parametric) locally and asymptotically most stringent test under a m-tuple $(f_1, \ldots, f_m)$ of FvML densities, and "correcting" it in such a way that the resulting test $\phi^{(n)}$ remains valid under the entire class of rotationally symmetric distributions. Let $\mathbf{X}_{i1}, \ldots, \mathbf{X}_{in_i}$, $i = 1, \ldots, m$, denote m independent samples, $\hat{\boldsymbol{\mu}}^{(n)}$ be an adequate root-n consistent estimator of the common value $\boldsymbol{\mu}$ of $\boldsymbol{\mu}_1, \ldots, \boldsymbol{\mu}_m$ under the null hypothesis, and put $\bar{\mathbf{X}}_i := n_i^{-1}\sum_{j=1}^{n_i}\mathbf{X}_{ij}$, $i = 1, \ldots, m$. The resulting locally and asymptotically most stringent pseudo-FvML test statistic corresponds to

$$Q^{(n)} = (p-1)\left(\sum_{i=1}^{m} n_i\hat{D}_i\,\bar{\mathbf{X}}_i'(\mathbf{I}_p - \hat{\boldsymbol{\mu}}^{(n)}(\hat{\boldsymbol{\mu}}^{(n)})')\bar{\mathbf{X}}_i - \sum_{i,j}^{m}\frac{n_in_j}{n}\hat{E}_i\hat{E}_j\,\bar{\mathbf{X}}_i'(\mathbf{I}_p - \hat{\boldsymbol{\mu}}^{(n)}(\hat{\boldsymbol{\mu}}^{(n)})')\bar{\mathbf{X}}_j\right)$$

where $\hat{D}_i$ and $\hat{E}_i$ are estimators of quantities which depend on the moments of the $\mathbf{X}_{ij}'\boldsymbol{\mu}$'s.

The rank-based version of $\phi_{\underline{f}}^{(n)}$ is obtained as follows. Since there exists a group of monotone transformations that generates the null hypothesis, we construct tests based on the maximal invariant associated with this group. Letting R_{ij} denote the rank of $\mathbf{X}'_{ij}\boldsymbol{\mu}$ among $\mathbf{X}'_{i1}\boldsymbol{\mu}, \ldots, \mathbf{X}'_{in_i}\boldsymbol{\mu}$, $i = 1, \ldots, m$, the maximal invariant is given by

$$\left(R_{11}, \ldots, R_{mn_m}, \mathbf{S}'_{\boldsymbol{\mu}}(\mathbf{X}_{11}), \ldots, \mathbf{S}'_{\boldsymbol{\mu}}(\mathbf{X}_{mn_m})\right)',$$

where $\mathbf{S}_{\boldsymbol{\mu}}(\mathbf{X}_{ij})$ represents the multivariate sign vector of $\mathbf{X}_{ij}$ with respect to the common value $\boldsymbol{\mu}$. The resulting tests are based on estimated versions of the spherical signs and ranks described above. They are asymptotically valid under any m-tuple of rotationally symmetric densities. This second approach, however, implies that for any given m-tuple $(f_1, \ldots, f_m)$ of rotationally symmetric densities (not necessarily FvML ones) one has to correctly choose the appropriate m-tuple $\underline{K} = (K_1, \ldots, K_m)$ of score functions to guarantee that the resulting test is asymptotically most stringent under $(f_1, \ldots, f_m)$. Letting $\hat{\boldsymbol{\mu}}^{(n)}$ still denote an adequate root-n consistent estimator of $\boldsymbol{\mu}$ and $\hat{R}_{ij}$ the rank of $\mathbf{X}'_{ij}\hat{\boldsymbol{\mu}}^{(n)}$ among $\mathbf{X}'_{i1}\hat{\boldsymbol{\mu}}^{(n)}, \ldots, \mathbf{X}'_{in_i}\hat{\boldsymbol{\mu}}^{(n)}$, put $\bar{\mathbf{U}}_i := n_i^{-1}\sum_{j=1}^{n_i}\mathbf{U}_{ij}$ with $\mathbf{U}_{ij} := K_i\left(\hat{R}_{ij}/(n_i+1)\right)\mathbf{S}_{\hat{\boldsymbol{\mu}}^{(n)}}(\mathbf{X}_{ij})$, $i = 1, \ldots, m$. A locally and asymptotically most stringent test then rejects the null hypothesis at asymptotic level α when the signed-rank test statistic

$$\underset{\sim}{Q}{}^{(n)}_{\underline{K}} := (p-1)\left(\sum_{i=1}^{m} n_i \hat{D}_{\underline{K},i}\bar{\mathbf{U}}'_i\bar{\mathbf{U}}_i - \sum_{i,j=1}^{m}\frac{n_i n_j}{n}\hat{E}_{\underline{K},i}\hat{E}_{\underline{K},j}\,\bar{\mathbf{U}}'_i\bar{\mathbf{U}}_j\right) > \chi^2_{(m-1)(p-1);1-\alpha}$$

where the $\hat{D}_{\underline{K},i}$'s and $\hat{E}_{\underline{K},i}$'s are estimators of cross-information quantities which depend on the choice of $\underline{K}$.

5.3.5 Application 3: Asymptotic power of tests of concentration

A further advantage of the Le Cam theory for spherical data described in Section 5.3.2 is that it can be used to calculate the power of testing procedures. We shall conclude this chapter by exhibiting how to calculate the power of tests for the concentration parameter of FvML distributions. These results were derived in Ley & Verdebout (2014*a*). Let the data points $\mathbf{X}_1, \ldots, \mathbf{X}_n$ be iid with common FvML density with location $\boldsymbol{\mu} \in \mathcal{S}^{p-1}$ and concentration $\kappa > 0$. We denote their joint distribution by $\mathrm{P}^{(n)}_{\boldsymbol{\vartheta}}$ with $\boldsymbol{\vartheta} := (\kappa, \boldsymbol{\mu}')' \in \mathbb{R}_0^+ \times \mathcal{S}^{p-1}$. Consider testing the null hypothesis $\mathcal{H}_0^{\kappa_0} : \kappa = \kappa_0$ for some fixed $\kappa_0 > 0$ against the alternative $\mathcal{H}_1^{\kappa_0} : \kappa \neq \kappa_0$. To derive efficient procedures for this problem and calculate their powers, Proposition 5.3.1 needs to be extended to take the concentration parameter into account. The following result holds for the FvML family.

Proposition 5.3.3 (Ley & Verdebout 2014*a*) *The family* $\left\{\mathrm{P}^{(n)}_{\boldsymbol{\vartheta}} \mid \boldsymbol{\vartheta} \in \mathbb{R}_0^+ \times \mathcal{S}^{p-1}\right\}$ *is ULAN. More precisely, for any sequence* $\boldsymbol{\vartheta}^{(n)} \in \mathbb{R}_0^+ \times \mathcal{S}^{p-1}$ *such that* $\boldsymbol{\vartheta}^{(n)} - \boldsymbol{\vartheta} =$

$O(n^{-1/2})$ and any bounded sequence $\boldsymbol{\tau}^{(n)} = (\tau_1^{(n)}, (\boldsymbol{\tau}_2^{(n)})') \in \mathbb{R} \times \mathbb{R}^p$ for which $\kappa^{(n)} + n^{-1/2}\tau_1^{(n)} > 0$ and $\tau_2^{(n)}$ satisfies condition (5.18), we have

$$\log\left(\frac{d\mathrm{P}^{(n)}_{\boldsymbol{\vartheta}^{(n)}+n^{-1/2}\boldsymbol{\tau}^{(n)}}}{d\mathrm{P}^{(n)}_{\boldsymbol{\vartheta}^{(n)}}}\right) = (\boldsymbol{\tau}^{(n)})'\boldsymbol{\Delta}^{(n)}_{\boldsymbol{\vartheta}^{(n)}} - \frac{1}{2}(\boldsymbol{\tau}^{(n)})'\boldsymbol{\Gamma}_{\boldsymbol{\vartheta}}\boldsymbol{\tau}^{(n)} + o_{\mathrm{P}}(1) \qquad (5.25)$$

and $\boldsymbol{\Delta}^{(n)}_{\boldsymbol{\vartheta}^{(n)}} \xrightarrow{\mathcal{D}} \mathcal{N}_{p+1}(\mathbf{0}, \boldsymbol{\Gamma}_{\boldsymbol{\vartheta}})$ under $\mathrm{P}^{(n)}_{\boldsymbol{\vartheta}^{(n)}}$ as $n \to \infty$. The central sequence

$$\boldsymbol{\Delta}^{(n)}_{\boldsymbol{\vartheta}} := \left(\left(\Delta^{(\mathrm{I})(n)}_{\boldsymbol{\vartheta}}\right), \left(\boldsymbol{\Delta}^{(\mathrm{II})(n)}_{\boldsymbol{\vartheta}}\right)'\right)'$$

is defined by

$$\Delta^{(\mathrm{I})(n)}_{\boldsymbol{\vartheta}} := n^{-1/2}\sum_{i=1}^{n}(\mathbf{X}_i'\boldsymbol{\mu} - A_p(\kappa))$$

with[3]

$$A_p(\kappa) = \frac{I_{p/2}(\kappa)}{I_{p/2-1}(\kappa)}$$

and

$$\boldsymbol{\Delta}^{(\mathrm{II})(n)}_{\boldsymbol{\vartheta}} := \kappa n^{-1/2}\sum_{i=1}^{n}(1 - (\mathbf{X}_i'\boldsymbol{\mu})^2)^{1/2}\mathbf{S}_{\boldsymbol{\mu}}(\mathbf{X}_i),$$

and the associated Fisher information is given by

$$\boldsymbol{\Gamma}_{\boldsymbol{\vartheta}} := \begin{pmatrix} 1 - \frac{p-1}{\kappa}A_p(\kappa) - (A_p(\kappa))^2 & 0 \\ 0 & \frac{\kappa^2 \mathcal{J}_p(\kappa)}{p-1}(\mathbf{I}_p - \boldsymbol{\mu}\boldsymbol{\mu}') \end{pmatrix},$$

where $\mathcal{J}_p(\kappa) := 1 - \mathrm{E}[(\mathbf{X}_i'\boldsymbol{\mu})^2]$ under $\mathrm{P}^{(n)}_{\boldsymbol{\vartheta}}$.

The reader will by now know how the ULAN property from Proposition 5.3.3 helps the construction of locally and asymptotically optimal (here, maximin) tests for $\mathcal{H}_0^{\kappa_0}$ against $\mathcal{H}_1^{\kappa_0}$. Since κ is the parameter of interest, locally and asymptotically maximin tests for $\mathcal{H}_0^{\kappa_0}$ can be built upon $\Delta^{(\mathrm{I})(n)}_{\boldsymbol{\vartheta}}$, the κ-part of the central sequence. The diagonal structure of the information matrix is attractive as it allows $\boldsymbol{\mu}$ to be replaced by a root-n consistent estimator $\hat{\boldsymbol{\mu}}^{(n)}$ without asymptotic effect on $\Delta^{(\mathrm{I})(n)}_{\boldsymbol{\vartheta}}$ (recall the asymptotic linearity property (5.4)). The resulting test rejects the null hypothesis at asymptotic level α when

$$Q^{(n)}_{\kappa_0} := \frac{\left(\sum_{i=1}^{n}\left(\mathbf{X}_i'\hat{\boldsymbol{\mu}}^{(n)} - A_p(\kappa_0)\right)\right)^2}{n(1 - \frac{p-1}{\kappa_0}A_p(\kappa_0) - (A_p(\kappa_0))^2)} > \chi^2_{1;1-\alpha}. \qquad (5.26)$$

[3]The quantity $A_p(\kappa)$ appearing in the κ-part of the central sequence is further discussed in Section 4.4 of Chapter 4.

The Third Le Cam Lemma (Proposition 5.2.2) is tailor-made to provide the asymptotic power of $Q_{\kappa_0}^{(n)}$ under local alternatives of the form $\bigcup_{\boldsymbol{\mu}\in\mathcal{S}^{p-1}}\mathrm{P}^{(n)}_{(\kappa_0+n^{-1/2}\tau_1^{(n)},\boldsymbol{\mu})}$. Writing $\tau_1 := \lim_{n\to\infty}\tau_1^{(n)}$, the power is given by semiparametricwhere $F_{\chi^2_\nu(y)}$ denotes the distribution function of the non-central chi-squared distribution with ν degrees of freedom and non-centrality parameter y.

5.4 Further reading

This chapter has described the extension of the Le Cam theory, especially of the ULAN property and ensuing inferential procedures, to the setting of rotationally symmetric data on unit hyperspheres. Chapter 6 contains further examples where this methodology is put to use. For the interested reader, we now mention papers that are related to the methods and statistics considered in the previous sections of the present chapter.

One-sample and multi-sample concentration problems

In Section 5.3.5 we explained how the Le Cam theory can be used to compute the power of a maximin test for concentration within the FvML family. This test was first derived in Watamori & Jupp (2005) as a score test. That same paper also proposed a score test for the multi-sample concentration homogeneity problem in FvML families, a test whose power was calculated in Ley & Verdebout (2014*a*). The multi-sample concentration homogeneity problem was also addressed in Verdebout (2015) from a semiparametric point of view by proposing rank-based tests for this problem.

Rank-based tests based on a different definition of ranks on spheres

Rank-based location tests for rotationally symmetric data on unit hyperspheres were first introduced in Neeman & Chang (2001), and a few years later Tsai & Sen (2007) built locally best rotation-invariant rank tests by invoking the First Le Cam Lemma (however without using the ULAN methodology). There is an important structural difference between the ranks they used and the ranks defined in Section 5.3.3. While the quantities R_i, $i = 1, \ldots, n$, are the ranks of $\mathbf{X}_i'\boldsymbol{\mu}$ amongst $\mathbf{X}_1'\boldsymbol{\mu}, \ldots, \mathbf{X}_n'\boldsymbol{\mu}$, Neeman & Chang (2001) and Tsai & Sen (2007) considered instead R_i^+ as the rank of $||\mathbf{X}_i - (\mathbf{X}_i'\boldsymbol{\mu})\boldsymbol{\mu}||$ amongst $||\mathbf{X}_1 - (\mathbf{X}_1'\boldsymbol{\mu})\boldsymbol{\mu}||, \ldots, ||\mathbf{X}_n - (\mathbf{X}_n'\boldsymbol{\mu})\boldsymbol{\mu}||$.

Recent results for tests of uniformity and symmetry

6.1 Introduction

In this chapter, we consider two fundamental types of test procedures on hyperspheres, namely tests for uniformity and for symmetry. The problem of testing uniformity on the unit hypersphere $\mathcal{S}^{p-1}$ is one of the oldest problems in directional statistics. It can be traced back to the discussion by Bernoulli (1735) on the solution to the problem of whether the closeness of the orbital planes of various planets arose by chance or not. Rayleigh (1880) was the first to study the resultant length of bivariate uniform unit vectors, and the first test of uniformity was proposed in Rayleigh (1919). Since then, the problem has attracted considerable attention. Kuiper (1960) studied Kolmogorov–Smirnov type tests for the circular case while Watson (1961) introduced Cramér–von Mises type tests. Ajne (1968) tested circular uniformity by comparing the number of observations in each semicircle with the expected value of $n/2$; this test was extended to $\mathcal{S}^{p-1}$ by Beran (1968). That same paper, together with Beran (1969), introduced a class of circular tests that are locally most powerful invariant against a specific non-parametric alternative, while Giné (1975) proposed Sobolev tests of uniformity on the unit hypersphere. Bingham (1974) constructed tests of uniformity of axial data by exploiting the idea that the sample scatter matrix should, under uniformity, be close to $\mathbf{I}_p/p$. Cordeiro & Ferrari (1991) modified the Rayleigh test in the circular case in order to improve the chi-squared approximation at the limit as $n \to \infty$. Jupp (2001) extended the results of Cordeiro & Ferrari (1991) to the hypersphere. We refer the reader to Sections 6.3, 8.3, 10.4.1, 10.7.1 and 10.8 of Mardia & Jupp (2000) for detailed information about most of the aforementioned tests.

Tests for symmetry on $\mathbb{R}^p$ have been widely studied in the past decades. Classical tests for (univariate or multivariate) symmetry include the Cramér–von Mises type test of Rothman & Woodrofe (1972), the elliptical symmetry test of Beran (1979*b*), the triples test of Randles et al. (1980), the runs test of McWilliams (1990) and the spherical symmetry test of Baringhaus (1991). This line of research is of continued interest as the recent contributions of Dyckerhoff et al. (2015), Einmahl & Gan (2016) and Partlett & Patil (2017) attest.

Testing for symmetry on directional supports has received considerably less attention in the literature. This can be seen from the very short Sections 8.2 and 10.7.2 in Mardia & Jupp (2000) dealing with this issue. The literature indeed is extremely scarce. Schach (1969) built tests for circular reflective symmetry about a fixed axis (see Section 2.2.2 for the notion of reflective symmetry), however these tests strongly resemble location tests. Mardia & Jupp (2000) suggest tackling the problem using suitably adapted versions of sign and one-sample Wilcoxon tests. An exploratory graphical tool to check if unimodal densities are reflectively symmetric is suggested in Section 4.2 of Fisher (1993). Jupp & Spurr (1983) proposed rank-based procedures for testing a different notion of symmetry on the circle, namely l-fold symmetry (which is not the aim of this chapter). On general hyperspheres, most tests for rotational symmetry boil down to parametric submodel tests, where a certain distribution is the only symmetric member of a larger family of distributions. Numerous likelihood ratio or Rao score tests have been proposed, see, e.g., Section 10.7.2 of Mardia & Jupp (2000). A very good account of such tests, referring also to new smooth goodness-of-fit tests, is given by Boulerice & Ducharme (1997) who provide generalized versions of the Beran (1979*a*) test based on spherical harmonics.

The recent growth in interest in non-symmetric distributions on the circle and sphere (see Sections 2.2 and 2.3) reflects an awareness of the need to go beyond the classical directional distributions founded on some notion of symmetry along with a need of tests for symmetry. In this chapter we highlight the new proposals of tests for symmetry from the year 2000 onwards.

6.1.1 Organization of the remainder of the chapter

We start, in Section 6.2, by describing recent advances concerning the Rayleigh test of uniformity. Section 6.3 contains results from Jupp (2008) related to Sobolev tests of uniformity, while tests based on random projections are presented in Section 6.4. In Section 6.5, we consider tests of uniformity in the presence of noisy data. Tests for reflective symmetry and rotational symmetry are addressed in Sections 6.6 and 6.7, respectively. Finally, we then turn our attention to location tests in the vicinity of uniformity, i.e., where the concentration parameter vanishes as a function of the sample size (Section 6.8).

6.2 Recent advances concerning the Rayleigh test of uniformity

Let $\mathbf{X}_1, \ldots, \mathbf{X}_n$ be a sample of iid random vectors on $\mathcal{S}^{p-1}$. The Rayleigh test (Rayleigh 1919) rejects the null hypothesis of uniformity $\mathcal{H}_0^{\text{unif}}$ for large values of

$$R_n = \frac{p}{n} \sum_{i,j=1}^{n} \mathbf{X}_i'\mathbf{X}_j = np\|\bar{\mathbf{X}}\|^2 \tag{6.1}$$

with $\bar{\mathbf{X}} = n^{-1}\sum_{i=1}^{n} \mathbf{X}_i$. Under $\mathcal{H}_0^{\text{unif}}$, the $\mathbf{X}_i$'s have mean zero and covariance matrix $\frac{1}{p}\mathbf{I}_p$ so that the multivariate Central Limit Theorem implies that, for any fixed p, the Rayleigh statistic R_n converges to χ_p^2 as $n \to \infty$. Therefore, Rayleigh's test rejects the null hypothesis, at asymptotic level α, whenever $R_n > \chi^2_{p;1-\alpha}$. A detailed account of the Rayleigh test is provided in Section 10.4.1 of Mardia & Jupp (2000). In particular, the Rayleigh test is the score test for testing uniformity against FvML distributions. It is also locally most powerful invariant against the same FvML alternatives as pointed out by Chikuse (2012). Jupp (2001) gave corrections to the Rayleigh test which improve the large-sample chi-squared approximation to the sampling distribution of R_n.

Very recently, Cutting et al. (2017*a*) showed that the Rayleigh test is also locally and asymptotically optimal in the Le Cam sense (see Chapter 5) against a family of rotationally symmetric distributions. Here we briefly outline the arguments that led to this result in the FvML case. Consider the reparametrization $\boldsymbol{\eta} = \kappa\boldsymbol{\mu}$, where $\kappa \geq 0$ and $\boldsymbol{\mu} \in \mathcal{S}^{p-1}$ are respectively the concentration and the location parameter of an FvML distribution. The null hypothesis of uniformity within the FvML family indexed by $\boldsymbol{\eta}$ can be represented by $\mathcal{H}_0 : \boldsymbol{\eta} = 0$. Consider now the log-likelihood ratio

$$\Lambda^{(n)} := \log\left(\frac{\mathrm{dP}^{(n)}_{\sqrt{\frac{p}{n}}\mathbf{e}^{(n)}}}{\mathrm{dP}^{(n)}_0}\right),$$

where $\mathrm{P}^{(n)}_{\boldsymbol{\eta}}$ denotes the joint distribution of the sample $\mathbf{X}_1, \ldots, \mathbf{X}_n$ of FvML random vectors with parameter $\boldsymbol{\eta}$, and $\mathbf{e}^{(n)}$ is a bounded sequence in $\mathbb{R}^p$. The reader will recognize here the formulations from Chapter 5. Like the log-likelihood ratios there, that can be written as Taylor expansions involving a central sequence and the Fisher information, the ratio $\Lambda^{(n)}$ can be expressed as

$$\Lambda^{(n)} = (\mathbf{e}^{(n)})'\mathbf{\Delta}^{(n)} - \frac{1}{2}\|\mathbf{e}^{(n)}\|^2 + o_{\mathrm{P}}(1)$$

as $n \to \infty$ under the null hypothesis of uniformity, with $\mathbf{\Delta}^{(n)} := \sqrt{\frac{p}{n}}\sum_{i=1}^{n}\mathbf{X}_i$ that is asymptotically normal with mean zero and covariance $\mathbf{I}_p$. It directly follows from the results of Section 5.2.3 that the test rejecting $\mathcal{H}_0$ at asymptotic level α when

$$(\mathbf{\Delta}^{(n)})'\mathbf{\Delta}^{(n)} > \chi^2_{p;1-\alpha}$$

is locally and asymptotically maximin against FvML alternatives. The alert reader will have noticed that $(\mathbf{\Delta}^{(n)})'\mathbf{\Delta}^{(n)}$ is R_n, hence the result applies specifically to the Rayleigh test.

As seen in Chapter 5, this enables us to calculate the asymptotic power of the Rayleigh test under local alternatives of the form $\mathrm{P}^{(n)}_{n^{-1/2}c^{(n)}\boldsymbol{\mu}}$ for some sequence $c^{(n)}$ with limiting value $c > 0$. That power is given by

$$1 - F_{\chi^2_p(c^2/p)}(\chi^2_{p;1-\alpha}),$$

and we refer the reader to the end of Section 5.3.5 for an explanation of the notation $F_{\chi^2_p(c^2/p)}(\chi^2_{p;1-\alpha})$.

6.3 Sobolev tests of uniformity

Despite its optimality properties against FvML distributions, the major drawback of the Rayleigh test is its non-consistency against alternatives for which $\mathrm{E}[\mathbf{X}] = 0$. This lack of consistency has motivated the construction of numerous alternative tests of uniformity, as mentioned in Section 6.1. Here we focus on *Sobolev tests of uniformity*, first proposed in Giné (1975). The idea underpinning the Giné procedures is to construct tests based on the eigenfunctions of the Laplacian operator acting on $\mathcal{S}^{p-1}$. Denoting by $\mathcal{E}_k$ (with dimension d_k) the space of eigenfunctions corresponding to the k-th non-zero eigenvalue of the Laplacian, there exists a well-defined mapping $\mathbf{t}_k : \mathcal{S}^{p-1} \to \mathcal{E}_k$ that can be written as

$$\mathbf{t}_k(\mathbf{x}) := \sum_{i=1}^{d_k} g_i(\mathbf{x})g_i,$$

where the g_i's form an orthonormal basis of $\mathcal{E}_k$. Letting $v_1, v_2, \ldots$ be a real sequence such that the series $\sum_{k=1}^{\infty} v_k^2 d_k$ is finite, the function

$$\mathbf{x} \mapsto \mathbf{t}(\mathbf{x}) := \sum_{k=1}^{\infty} v_k \mathbf{t}_k(\mathbf{x}) \tag{6.2}$$

is a mapping from $\mathcal{S}^{p-1}$ to the Hilbert space $L^2(\mathcal{S}^{p-1}, d\mathrm{P}_\mathrm{u})$ of measurable functions f for which

$$\int_{\mathcal{S}^{p-1}} f^2(\mathbf{x})\, d\mathrm{P}_\mathrm{u}(\mathbf{x}) < \infty,$$

where $d\mathrm{P}_\mathrm{u}$ is the uniform measure on $\mathcal{S}^{p-1}$. Letting $\mathbf{X}_1, \ldots, \mathbf{X}_n$ be an iid sample of unit random vectors, the Giné test rejects the null hypothesis of uniformity $\mathcal{H}_0^{\mathrm{unif}}$ for large values of

$$S_n := n^{-1} \left\| \sum_{i=1}^{n} \mathbf{t}(\mathbf{X}_i) \right\|_{L^2}^2 = n^{-1} \sum_{i,j=1}^{n} \langle \mathbf{t}(\mathbf{X}_i), \mathbf{t}(\mathbf{X}_j) \rangle,$$

where $\langle \cdot, \cdot \rangle$ denotes the inner product on $L^2(\mathcal{S}^{p-1}, d\mathrm{P}_\mathrm{u})$ defined as

$$\langle f, g \rangle := \int_{\mathcal{S}^{p-1}} f(\mathbf{x}) g(\mathbf{x}) d\mathrm{P}_\mathrm{u}(\mathbf{x}).$$

Closed forms of $\mathbf{t}$ for $\mathcal{S}^1$ and $\mathcal{S}^2$ are provided in Section 6 of Giné (1975).

This construction was revisited in recent years, when Jupp (2008) constructed new Sobolev-type tests. More precisely, Jupp (2008) started by showing that the score test of uniformity against the exponential model proposed in Beran (1979*a*) is also of the form (6.2) and rejects uniformity for large values of

$$S_k := n^{-1} \left\| \sum_{i=1}^{n} \mathbf{t}_{(k)}(\mathbf{X}_i) \right\|_{L^2}^2,$$

where $\mathbf{t}_{(k)}$ corresponds to $\mathbf{t}$ computed with the weights $v_j = 1$ for $j \leq k$ and $v_j = 0$ for $j > k$. The major problem with tests based on S_k is the selection of k. Jupp (2008) suggested a data-driven selection of k based on a modification of the Bayesian Information Criterion. Using the penalized score statistic

$$B_S(k) := S_k - \left(\sum_{i=1}^{k} d_i \right) \log n,$$

where the d_i's again denote the dimensions of the spaces of eigenfunctions, the proposed estimator $\hat{k}$ of k is

$$\hat{k} := \inf_{k \in \mathbb{N}} \left\{ B_S(k) = \sup_{m \in \mathbb{N}} B_S(m) \right\}.$$

We remark that the infimum of the empty set is ∞. This choice yields the following properties:

(i) $\hat{k}$ is almost surely finite in the absolutely continuous case; that is, $\mathrm{P}[\hat{k} = \infty] = 0$ when the sample size n is larger than 3;

(ii) under the null hypothesis of uniformity, $\hat{k}$ converges in probability to one.

Under the null hypothesis of uniformity, the test statistic $S_{\hat{k}}$ is asymptotically chi-squared with d_1 degrees of freedom, a result closely related to (ii) above. The resulting test that rejects $\mathcal{H}_0^{\text{unif}}$ at asymptotic level α when $S_{\hat{k}} > \chi^2_{d_1;1-\alpha}$ is universally consistent, unlike the Rayleigh test.

6.4 Uniformity tests based on random projections

A class of non-parametric tests based on random projections for testing uniformity on $\mathcal{S}^{p-1}$ was introduced by Cuesta-Albertos et al. (2009). Consider a sample $\mathbf{X}_1, \ldots, \mathbf{X}_n$ of iid random vectors on $\mathcal{S}^{p-1}$. Letting $\mathbf{U}$ denote a uniformly distributed vector on $\mathcal{S}^{p-1}$ that is independent of the $\mathbf{X}_i$'s, the common distribution of the projections

$$Y_1 = \mathbf{X}_1'\mathbf{U}, \ldots, Y_n = \mathbf{X}_n'\mathbf{U}$$

uniquely determines (with probability one) the distribution of the $\mathbf{X}_i$'s. We denote by F_0 the common distribution of the Y_i's when the $\mathbf{X}_i$'s are uniform. Closed forms for F_0 for the important cases $p = 2$ and $p = 3$ are available. For example, when $p = 3$, F_0 is the uniform distribution on the unit interval $[-1, 1]$. The random projection test can be summarized by the following two steps:

(i) Compute the projections $Y_1, \ldots, Y_n$.

(ii) Letting F_n denote the empirical distribution function of the Y_i's, the null hypothesis of uniformity is rejected for large values of

$$T_n = \sup_{x \in [-1,1]} |F_n(x) - F_0(x)|.$$

Note that T_n is a Kolmogorov–Smirnov test statistic for F_0. Consequently, the critical values of T_n are obtained via those of a traditional Kolmogorov–Smirnov test.

The tests proposed by Cuesta-Albertos et al. (2009) are not based on a single projection $\mathbf{U}$ (and therefore a single p-value or critical value). Instead they consider N random projections and use a "Bonferroni type" procedure to achieve the correct nominal level. The conclusion of the simulation study performed in Cuesta-Albertos et al. (2009) is that in terms of empirical level/power the overall performances of tests based on random projections is quite satisfactory.

6.5 Testing for uniformity with noisy data

We shall now deviate slightly from the uniformity tests of the previous sections and consider the more complex situation in which we seek to test uniformity in the presence of noisy data. Such data typically arise in astronomy where the incoming directions of cosmic rays are very likely to be perturbed by galactic and intergalactic fields.

Consider two independent sequences $\mathbf{X}_1, \ldots, \mathbf{X}_n$ and $\epsilon_1, \ldots, \epsilon_n$ where the $\mathbf{X}_i$'s are iid on the sphere $\mathcal{S}^2$ and the ϵ_i's are iid on $\mathcal{SO}(3)$, the rotation group in $\mathbb{R}^3$. Assume that the observed sequence on $\mathcal{S}^2$ is of the form

$$\mathbf{X}_1^* = \epsilon_1 \mathbf{X}_1, \ldots, \mathbf{X}_n^* = \epsilon_n \mathbf{X}_n.$$

If both sequences $\mathbf{X}_1, \ldots, \mathbf{X}_n$ and $\epsilon_1, \ldots, \epsilon_n$ are absolutely continuous with respect to the uniform measure on $\mathcal{S}^2$ and the Haar measure on $\mathcal{SO}(3)$, respectively, then the common density $f_{\mathbf{X}^*}$ of the $\mathbf{X}_i^*$'s is the convolution product

$$f_{\mathbf{X}^*}(\mathbf{x}) = f_\epsilon \star f(\mathbf{x}) := \int_{\mathcal{SO}(3)} f_\epsilon(\mathbf{R}) f(\mathbf{R}^{-1}\mathbf{x}) d\mathbf{R}$$

of the common density f of the $\mathbf{X}_i$'s with the common density f_ϵ of the ϵ_i's. Lacour & Pham Ngoc (2014) and Kim et al. (2016) considered the problem of testing the null hypothesis $\mathcal{H}_0 : f = f_0$ with f_0 the uniform density on $\mathcal{S}^2$ in a setup in which the density f_ϵ of the noise is assumed to be known. The alternative $\mathcal{H}_1 = \mathcal{H}(\mathcal{F}, \delta, M)$ consists in a set of densities

$$\mathcal{H}(\mathcal{F}, \delta, M) := \{f \in \mathcal{F}, \|f - f_0\|_{L^2} \geq M\delta\}$$

where M is a constant and δ is referred to as the separation rate. In Lacour & Pham Ngoc (2014) the set $\mathcal{F}$ is a Sobolev class on the unit sphere with smoothness s defined as follows. Letting $C^\infty(\mathcal{S}^2)$ denote the space of infinitely continuously differentiable functions on $\mathcal{S}^2$, the Sobolev norm $\|\cdot\|_{W_s}$ of a function f is defined as

$$\|f\|_{W_s}^2 := \sum_{l=0}^{\infty} \sum_{m=-l}^{l} (1 + l(l+1))^s \hat{f}_l^2,$$

where the $\hat{f}_l$'s are the components of the *spherical Fourier transform* of f; see Lacour & Pham Ngoc (2014) for details. The Sobolev norm of any probability density f on $\mathcal{S}^2$ is such that $\|f\|^2_{W_s} \geq (4\pi)^{-1}$ with equality for the uniform distribution. The class $\mathcal{F}$ then consists of those densities f that belong to the Sobolev space $W_s(\mathcal{S}^2)$ (the completion of $C^\infty(\mathcal{S}^2)$ with respect to the s-norm) and satisfy

$$\|f\|^2_{W_s} \leq \frac{1}{4\pi} + c$$

for some constant $c > 0$. Let $\delta = \delta_n$ be a sequence indexed by n. If f_ϵ belongs to some class with regularity ν (assumed to be known), an adaptive procedure (that does not require the specification of s) cannot have a faster separation rate than

$$\delta_n = (n/\sqrt{\log\log n})^{-2s/(2(s+\nu)+1)}.$$

Lacour & Pham Ngoc (2014) proposed a test that achieves this rate.

6.6 Tests of reflective symmetry on the circle

Let $\Theta_1, \ldots, \Theta_n$ be iid random angles on $[-\pi, \pi)$. Ley & Verdebout (2014*b*) built semiparametric tests for reflective symmetry about a fixed center $\mu \in [-\pi, \pi)$ that are designed to be efficient against *k-sine-skewed alternatives*

$$\theta \mapsto f_0(\theta - \mu)(1 + \lambda \sin(k(\theta - \mu))),$$

a special family of the densities defined in Section 2.2.5. Here f_0 is reflectively symmetric about μ, and $\lambda \in (-1, 1)$ plays the role of skewness parameter. Hence the null hypothesis of reflective symmetry corresponds to $\mathcal{H}_0^{\rm sym} : \lambda = 0$. This null hypothesis is semiparametric, as f_0 is not specified. Following the Le Cam methodology described in Chapter 5, the semiparametrically efficient test for $\mathcal{H}_0^{\rm sym}$ rejects the null hypothesis for too large or small values of

$$Q_k^{(n)} := \frac{n^{-1/2}\sum_{i=1}^n \sin(k(\Theta_i - \mu))}{\left(n^{-1}\sum_{i=1}^n \sin^2(k(\Theta_i - \mu))\right)^{1/2}}. \tag{6.3}$$

This test is obtained by means of a studentization of the optimal f_0-parametric test.[1] More precisely, the test $\phi_k^{(n)}$ that rejects $\mathcal{H}_0^{\rm sym}$ at the asymptotic level α when

[1]For more details on its construction, we refer the reader to Section 6.7 where efficient semiparametric tests for rotational symmetry on hyperspheres against skew-rotationally-symmetric alternatives, again under specified location, are derived.

$|Q_k^{(n)}|$ exceeds the $\alpha/2$-upper quantile of a standard normal distribution is locally and asymptotically most powerful in the Le Cam sense against k-sine-skewed alternatives. Quite remarkably, $\phi_k^{(n)}$ is not only valid under any f_0, but also the optimal test against *any* k-sine-skewed f_0 alternative. Such uniform optimality occurs very rarely. The test with $k = 1$ is optimal against the sine-skewed alternatives of Abe & Pewsey (2011*a*), while the test for $k = 2$ coincides with the b_2^* test proposed a decade earlier by Pewsey (2004).

These powerful properties are lost when the location parameter μ is not assumed to be known. This situation was considered by Pewsey (2002) who built a test rejecting reflective symmetry for large absolute values of

$$T^{(n)} := \frac{n^{-1/2}\sum_{i=1}^{n}\sin(2(\Theta_i - \hat{\mu}))}{\sqrt{\widehat{\mathrm{Var}}[\sin(2(\Theta_i - \hat{\mu}))]}},$$

where $\hat{\mu}$ is the method of moments estimator of μ and $\widehat{\mathrm{Var}}[\sin(2(\Theta_i - \hat{\mu}))]$ is a consistent estimator (under the null hypothesis) of $\mathrm{Var}[\sin(2(\Theta_i - \hat{\mu}))]$. Pewsey (2002) shows that $T^{(n)}$ is asymptotically normal under the null hypothesis. This omnibus test is based on the sample second sine moment about $\hat{\mu}$, which is a measure of skewness popularized by Batschelet (1965).

6.7 Tests of rotational symmetry on hyperspheres

The higher-dimensional analogue of reflective symmetry on S^1 is rotational symmetry on $\mathcal{S}^{p-1}$ for $p > 2$, see Section 2.3.2. Similarly, the skew-rotationally-symmetric densities of Section 2.3.3 are natural extensions of the sine-skewed distributions on the circle. It is hence not surprising that efficient tests for rotational symmetry about a fixed direction $\boldsymbol{\mu} \in \mathcal{S}^{p-1}$ were constructed against alternatives of the form (2.23). Ley & Verdebout (2017) proposed a most efficient test for the null hypothesis $\mathcal{H}_0^{\mathrm{sym}} : \boldsymbol{\lambda} = \mathbf{0}$ against the alternative that $\boldsymbol{\lambda} \in \mathbb{R}^{p-1} \setminus \{\mathbf{0}\}$. We shall outline their construction in what follows.

Consider a sample of iid observations $\mathbf{X}_1, \ldots, \mathbf{X}_n$ on $\mathcal{S}^{p-1}$ with common density (2.23). For any rotationally symmetric density with angular function f_a and any skewing function Π, denote by $\mathrm{P}^{(n)}_{\boldsymbol{\vartheta};f_a,\Pi}$, with $\boldsymbol{\vartheta} := (\boldsymbol{\mu}', \boldsymbol{\lambda}')' \in \mathcal{S}^{p-1} \times \mathbb{R}^{p-1}$, the joint distribution of $\mathbf{X}_1, \ldots, \mathbf{X}_n$. Since, for $\boldsymbol{\lambda} = \mathbf{0}$, the resulting density does not depend on Π, we drop the index Π and simply write $\mathrm{P}^{(n)}_{\boldsymbol{\vartheta}_0;f_a}$ at $\boldsymbol{\vartheta}_0 := (\boldsymbol{\mu}', \mathbf{0}')'$. Letting $(\mathbf{e}^{(n)}, \mathbf{d}^{(n)})$ be bounded sequences in $\mathbb{R}^p \times \mathbb{R}^{p-1}$ such that

$\boldsymbol{\mu} + n^{-1/2}\mathbf{e}^{(n)} \in \mathcal{S}^{p-1}$, the first step consists in showing that the log-likelihood ratio

$$\log \frac{\mathrm{dP}^{(n)}_{((\boldsymbol{\mu}+n^{-1/2}\mathbf{e}^{(n)})',n^{-1/2}(\mathbf{d}^{(n)})')';f_a,\Pi}}{\mathrm{dP}^{(n)}_{(\boldsymbol{\mu}',\mathbf{0}')';f_a}}$$

admits a ULAN decomposition as seen in Chapter 5. We leave this task to the reader. It then follows from the general results of Chapter 5 that the locally and asymptotically optimal test for $\mathcal{H}^{\mathrm{sym}}_{0;f_a} : \boldsymbol{\lambda} = 0$ against $\mathcal{H}_{1;f_a} : \boldsymbol{\lambda} \neq 0$ is based on the central sequence

$$\boldsymbol{\Delta}^{(n)}_{\Pi;2}(\boldsymbol{\mu}) \;=\; 2\Pi'(0)n^{1/2}\boldsymbol{\Upsilon}'_{\boldsymbol{\mu}}\bar{\mathbf{X}},$$

where Π' is the derivative of Π. Under the f_a-parametric null hypothesis, this statistic is asymptotically normal with mean zero and covariance matrix

$$\boldsymbol{\Gamma}_{f_a,\Pi;\boldsymbol{\vartheta}_0} = \frac{4(\Pi'(0))^2\mathcal{A}_p(f_a)}{p-1}\mathbf{I}_{p-1}$$

with $\mathcal{A}_p(f_a) := 1 - \mathrm{E}_{f_a}[(\mathbf{X}_1'\boldsymbol{\mu})^2] < \infty$. Consequently, the locally and asymptotically optimal f_a-parametric procedure $\phi^{(n)}_{f_a}$ consists in rejecting the parametric null hypothesis at asymptotic level α whenever

$$\begin{aligned} T^{(n)}_{f_a}(\boldsymbol{\mu}) \;&:=\; \left(\boldsymbol{\Delta}^{(n)}_{\Pi;2}(\boldsymbol{\mu})\right)'(\boldsymbol{\Gamma}_{f_a,\Pi;\boldsymbol{\vartheta}_0})^{-}\boldsymbol{\Delta}^{(n)}_{\Pi;2}(\boldsymbol{\mu}) \\ &=\; \frac{n(p-1)}{\mathcal{A}_p(f_a)}\bar{\mathbf{X}}'(\mathbf{I} - \boldsymbol{\mu}\boldsymbol{\mu}')\bar{\mathbf{X}} \end{aligned}$$

exceeds $\chi^2_{p-1;1-\alpha}$. Here $\mathbf{A}^-$ denotes the Moore–Penrose pseudo-inverse of the matrix $\mathbf{A}$.

In order to render this f_a-parametric test semiparametric, and hence tackle the general hypothesis $\mathcal{H}^{\mathrm{sym}}_0$, one needs to estimate the quantity $\mathcal{A}_p(f_a)$. This can be achieved using the consistent (under any f_a) estimator $\hat{\mathcal{A}}_p := 1 - n^{-1}\sum_{i=1}^n(\mathbf{X}_i'\boldsymbol{\mu})^2$, so that the resulting testing procedure $\phi^{(n)}$ rejects $\mathcal{H}^{\mathrm{sym}}_0$ at asymptotic level α whenever

$$T^{(n)} \;=\; n(p-1)\hat{\mathcal{A}}_p^{-1}\bar{\mathbf{X}}'(\mathbf{I} - \boldsymbol{\mu}\boldsymbol{\mu}')\bar{\mathbf{X}} \tag{6.4}$$

exceeds $\chi^2_{p-1;1-\alpha}$. Note that $T^{(n)}$ does not depend on f_a: the test $\phi^{(n)}$ is hence not only valid under any f_a, but also uniformly the most efficient test, mimicking thus the uniformly optimal test for reflective symmetry described in Section 6.6. To the best of our knowledge, this test is the first test for rotational symmetry. Quite interestingly, the test statistic (6.4) coincides with the statistic used in Watson (1983) to address the directional location problem under rotational symmetry. Hence the Watson test, well known as efficient for the location problem, also happens to be an efficient test for rotational symmetry against skew-rotationally-symmetric alternatives under specified location $\boldsymbol{\mu}$.

Testing rotational symmetry against skew-rotationally-symmetric alternatives under unspecified location $\boldsymbol{\mu}$ is more delicate. First, the associated parametric test statistics will involve the angular function f_a under which they are built, preventing uniformly optimal tests. Second, the need to estimate $\boldsymbol{\mu}$ requires the consideration of the Fisher information matrix associated with $\boldsymbol{\mu}$ and $\boldsymbol{\lambda}$. This matrix can become singular. Ley & Verdebout (2017) indeed proved that this matrix is of lower rank in the vicinity of $\boldsymbol{\lambda} = \mathbf{0}$ if and only if f_a is the FvML angular function. This Fisher information singularity result very much resembles a similar result in $\mathbb{R}^p$, where the only base symmetric density leading to singular information matrices in skew-symmetric models is the multivariate normal, as established in Ley & Paindaveine (2010*b*) and Hallin & Ley (2012*b*).

6.8 Testing for spherical location in the vicinity of the uniform distribution

The topic of this section differs from the main theme of the present chapter. Consider the following situation: we observe spherical data that look uniformly distributed, but traditional tests will reject uniformity. Testing for the center of such (non-uniform) data is possible albeit very delicate, given the very low concentration around that center. For example, the Rayleigh test applied to the incoming directions of cosmic rays in Figure 6.1 rejects the null hypothesis of uniformity. However, a quick inspection of Figure 6.1 reveals that even if the data are not uniform they are very lowly concentrated. We shall now describe how to deal with the location testing problem $\mathcal{H}_0 : \boldsymbol{\mu} = \boldsymbol{\mu}_0$ against $\mathcal{H}_1 : \boldsymbol{\mu} \neq \boldsymbol{\mu}_0$, where $\boldsymbol{\mu}_0 \in \mathcal{S}^{p-1}$ is fixed, in the vicinity of the uniform distribution. This complicated version of a classical problem has been very recently addressed in Paindaveine & Verdebout (2017).

Let $\mathbf{X}_{ni}, i = 1, \ldots, n, n = 1, 2, \ldots,$ be a triangular array of observations where, for each n, $\mathbf{X}_{n1}, \ldots, \mathbf{X}_{nn}$ are independent and identically distributed with common rotationally symmetric density of the form $\mathbf{x} \mapsto c_{f_a,\kappa_n} f_a(\kappa_n \boldsymbol{\mu}_n' \mathbf{x})$ with angular function f_a, location $\boldsymbol{\mu}_n \in \mathcal{S}^{p-1}$ and concentration $\kappa_n \geq 0$ for all n. The

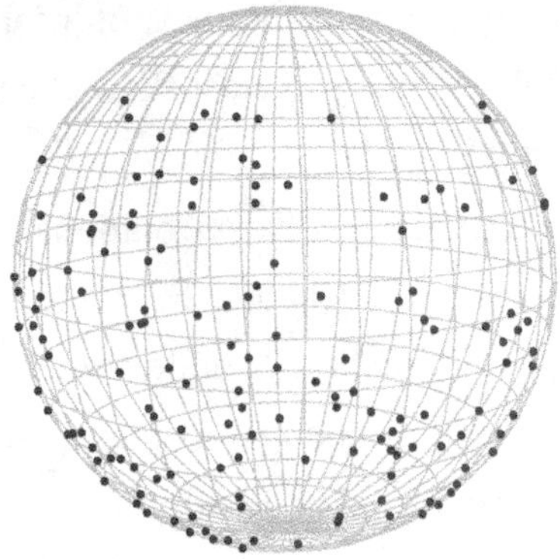

Figure 6.1: Representation of $n = 148$ measurements of incoming directions of cosmic rays. The data come from Toyoda et al. (1965).

associated sequence of joint distributions will be denoted as $\mathrm{P}^{(n)}_{\boldsymbol{\mu}_n,\kappa_n,f_a}$. A sequence $\mathrm{P}^{(n)}_{\boldsymbol{\mu}_n,\kappa_n,f_a}$ is said to be in a η_n-neighborhood of uniformity with locality parameter ξ if $\kappa_n = \sqrt{p}\xi\eta_n + o_{\mathrm{P}}(1)$ as $n \to \infty$. Therefore the sequence η_n will determine "how close" the sequence of hypotheses $\mathrm{P}^{(n)}_{\boldsymbol{\mu}_n,\kappa_n,f_a}$ is to the uniform case. We call settings where $\eta_n \to 0$ when $n \to \infty$ "shrinking to uniformity (SU)".

Within such sequences $\mathrm{P}^{(n)}_{\boldsymbol{\mu}_n,\kappa_n,f_a}$, Paindaveine & Verdebout (2017) considered the sequence of testing problems $\mathcal{H}_0^{(n)} : \boldsymbol{\mu}_n = \boldsymbol{\mu}_0$ and compared the performances of two traditional semiparametric tests that are based on the sample average $\bar{\mathbf{X}}_n := \frac{1}{n}\sum_{i=1}^n \mathbf{X}_{ni}$. More precisely, their goal was to check if these tests remain valid in the special setting where the concentration can tend to zero as a function of the sample size. The first test under investigation is the Watson test ϕ_n^W built upon W_n defined in (7.4). The second test is the Wald test ϕ_n^S (Hayakawa & Puri 1985, Hayakawa 1990) based on the test statistic

$$S_n := \frac{n(p-1)(\boldsymbol{\mu}_0'\bar{\mathbf{X}}_n)^2 \left(\frac{\bar{\mathbf{X}}_n}{||\bar{\mathbf{X}}_n||}\right)' (\mathbf{I}_p - \boldsymbol{\mu}_0\boldsymbol{\mu}_0') \frac{\bar{\mathbf{X}}_n}{||\bar{\mathbf{X}}_n||}}{1 - \frac{1}{n}\sum_{i=1}^n (\mathbf{X}_{ni}'\boldsymbol{\mu}_0)^2}.$$

Both tests reject a concentration-fixed null hypothesis $\mathcal{H}_0 : \boldsymbol{\mu} = \boldsymbol{\mu}_0$ against $\mathcal{H}_1 : \boldsymbol{\mu} \neq \boldsymbol{\mu}_0$ at asymptotic level α whenever their respective test statistics exceed $\chi^2_{p-1;1-\alpha}$.

Unsurprisingly, the two tests ϕ_n^W and ϕ_n^S are asymptotically equivalent away from uniformity (when $\eta_n = O(1)$) in the sense that $S_n - W_n$ is $o_{\mathrm{P}}(1)$ as $n \to \infty$

under $\mathrm{P}^{(n)}_{\mu_n,\kappa_n,f_a}$. However, they exhibit very different asymptotic behaviors under $\eta_n = o(1)$-neighborhoods. More precisely, W_n converges weakly to a chi-squared random variable with $p-1$ degrees of freedom in any $\eta_n = o(1)$-neighborhood as $n \to \infty$. On the contrary, S_n is no longer asymptotically chi-squared under η_n-neighborhoods of the form $\eta_n n \to 1$ as $n \to \infty$. We refer the reader to Paindaveine & Verdebout (2017) for the exact expression of the limiting distribution of S_n in such settings. We thus conclude that the Watson test ϕ_n^W is robust to SU settings while the Wald test ϕ_n^S is not.

It would be desirable that the robustness of the Watson test ϕ_n^W is not obtained at the expense of efficiency. Therefore it is essential to also study the asymptotic behavior of ϕ_n^W under $\mathrm{P}^{(n)}_{\mu_n,\kappa_n,f_a}$ with $\mu_n := \mu_0 + \nu_n \tau^{(n)}$ as $n \to \infty$ and for some bounded sequence $\tau^{(n)} \in \mathbb{R}^p$ such that $\mu_n \in \mathcal{S}^{p-1}$ for all n. Paindaveine & Verdebout (2017) obtained expressions for the local power of ϕ_n^W under various η_n-neighborhoods and showed that

- when $\eta_n = 1$, the Watson test is optimal at the FvML distribution only;
- when η_n is $o(1)$ with $\eta_n\sqrt{n} \to \infty$ as $n \to \infty$, the Watson test is optimal uniformly-in-f_a;
- when $\eta_n\sqrt{n} \to 1$ as $n \to \infty$, the Watson test is optimal uniformly-in-f_a but only locally-in-$\tau^{(n)}$;
- when $\eta_n\sqrt{n} \to 0$ as $n \to \infty$, the Watson test is optimal uniformly-in-f_a, but in a degenerate way.

As a conclusion, the Watson test ϕ_n^W remains an extremely competitive test in semi-parametric rotationally symmetric models when one has to deal with distributions that are close to uniformity.

6.9 Further reading

Uniformity tests

We have already mentioned in Section 6.1 various other tests for uniformity, and referred to the relevant references. Sobolev-like tests, as introduced in Giné (1975), have been revived in recent years, as described in Section 6.3. A review of data-driven versions of Sobolev tests for uniformity on Riemannian manifolds is provided in Jupp (2009). The same paper also discusses a variant of such tests for product manifolds. Sobolev tests for uniformity were also investigated in Bourguin et al. (2016), under the additional difficulties of high-frequency data and data available only on a portion of the sphere.

Another universally consistent test for uniformity was suggested in Pycke (2007). The author considers a U-statistic based on the geometric mean of distances between observations. That test statistic can be decomposed into various statistics that are asymptotically independent.

Universally consistent tests are not designed to be particularly powerful against a specific alternative. Most efficient tests can be built by nesting the uniform distribution into a family of non-uniform distributions and then considering likelihood ratio or Rao score tests. It is well known, and reestablished in Section 6.2, that the Rayleigh test is the most efficient test against FvML alternatives. The Beran-type tests are also locally most powerful invariant tests against a certain type of alternative. In recent years, Su & Wu (2011) constructed a test of uniformity against embedding smooth alternatives based on spherical harmonics.

Finally, it is interesting to mention recent power comparison studies between distinct uniformity tests. Figueiredo (2007) compared the Rayleigh, Ajne and Giné tests against FvML alternatives, while Figueiredo & Gomes (2003) studied the power of the Bingham and Giné tests against Watson distributions. These studies complement earlier findings presented, for example, in Stephens (1969) and Diggle et al. (1985).

Symmetry tests

Besides the b_2^* test mentioned in Section 6.6, Pewsey (2004) proposed three further tests of reflective symmetry about a known median direction. All three are circular versions of tests for symmetry on the real line: the one-sample Wilcoxon test, the runs test of McWilliams (1990) and the modified runs test of Modarres & Gastwirth (1996). We refer to Pewsey (2004) for a comparison of the performance of those tests by means of a Monte Carlo simulation study.

High-dimensional directional statistics

7.1 Introduction

7.1.1 High-dimensional techniques in $\mathbb{R}^p$

Modern computer tools allow the collection and the storage of massive datasets, a recent phenomenon commonly referred to as "big data". The treatment and analysis of these increasingly vast and complex datasets is nowadays one of the biggest challenges for statisticians and data analysts. Consequently, in the last decade there has been a huge activity related to high-dimensional problems. In particular, classification and dimension reduction problems have been considered in many recent papers. Classification techniques such as linear discriminant analysis, support vector machines, tree classifiers and nearest neighbor classifiers are still well used and studied techniques (see, e.g., Paindaveine & Van Bever 2015 and Scornet et al. 2015). In high-dimensional situations, however, those classical techniques tend to perform poorly, as pointed out by Bickel & Levina (2004). As a consequence, new high-dimensional classification methods have recently been brought to the statistical community: Guo et al. (2007) proposed regularized discriminant analysis, Witten & Tibshirani (2011) provided methods based on penalized linear discriminant analysis, while methods based on random projections were investigated in Cannings & Samworth (2017).

Principal Component Analysis (PCA) has also attracted much attention in the high-dimensional case. Consider $\mathbf{Z}_1, \ldots, \mathbf{Z}_n$ a random sample of iid centered random vectors taking values in $\mathbb{R}^p$. Classical principal components (PCs) are obtained by projecting the $\mathbf{Z}_i$'s along the eigenvectors of the sample covariance matrix $\mathbf{S} := n^{-1} \sum_{i=1}^{n} \mathbf{Z}_i \mathbf{Z}_i'$. The situation changes completely in the high-dimensional case: without any further assumptions, the eigenvectors of $\mathbf{S}$ are no longer consistent when the dimension p becomes arbitrarily large. In particular, Johnstone (2001) showed that letting $\hat{\mathbf{v}}$ and $\mathbf{v}$ respectively be the eigenvectors associated with the largest eigenvalue of $\mathbf{S}$ and $\mathbf{\Sigma}$, the unknown common covariance matrix of the observations, the angle between $\hat{\mathbf{v}}$ and $\mathbf{v}$ may not converge to 0 when p is of the form $p_n = O(n)$. As a consequence, several papers have been devoted to estimating PCs in sparse high-dimensional models, such as Cai, Ma & Wu (2013), Croux et al. (2013) and Han & Liu (2014), among many others.

Test procedures related to high-dimensional PCA have also been widely studied over the past years. In particular, many papers have addressed the problem of testing sphericity or unit covariance against *spiked covariance matrices* of the form

$$\sigma(\mathbf{I}_p + \lambda \mathbf{v}\mathbf{v}'), \tag{7.1}$$

where $\mathbf{I}_p$ is the identity matrix, $\sigma > 0, \lambda > 0$ and $\mathbf{v}$ is a unit vector. In particular, Ledoit & Wolf (2002), Birke & Dette (2005), Bai et al. (2009) and Chen et al. (2010) proposed different tests for this problem in high-dimensional models. Berthet & Rigollet (2013) tackled the problem when $\mathbf{v}$ is sparse while Onatski et al. (2013) computed high-dimensional powers of many different tests for this problem when $p = p_n = O(n)$.

High-dimensional tests, in general, have been in the spotlight recently. One natural and important challenge is to check if test procedures based on the classical n-asymptotics approximations remain valid in the new (n, p)-asymptotics framework where p can potentially be larger than n. Checking this validity is a nontrivial issue, which either leads to the confirmation of existing tests or reveals the need to modify them. Under the impetus of the influential paper by Ledoit & Wolf (2002), a large number of problems have been investigated, such as one- and multi-sample location tests (Chen & Qin 2010, Srivastava & Kubokawa 2013), one- and multi-sample covariance/scatter matrix tests (Chen et al. 2010, Li & Chen 2012) or likelihood ratio tests (Jiang & Yang 2013).

7.1.2 Organization of the remainder of the chapter

We start the chapter by describing in Section 7.2 distributions on $\mathcal{S}^{p-1}$ when p becomes very large. Having familiarized the reader with the difficulties of high-dimensional directional statistics, we then show how classical inferential problems have been solved for these settings: uniformity tests (Section 7.3), location tests (Section 7.4), and concentration tests (Section 7.5). We conclude the chapter with a discussion of the directional analogue of PCA, namely Principal Nested Spheres (Section 7.6).

7.2 Distributions on high-dimensional spheres

The present section is meant to complement Chapter 2 as we introduce models for a random vector $\mathbf{X}$ distributed on $\mathcal{S}^{p-1}$ with p large.

We start by showing, via some heuristics, that the unit norm constraint can arise naturally in high-dimensional situations. Consider a Gaussian random vector $\mathbf{Z}$ with mean zero and covariance matrix $p^{-1}\mathbf{I}_p$ (meaning that $\mathbf{Z}$ is spherically symmetric). Consequently, $p\|\mathbf{Z}\|^2$ follows a chi-squared distribution with p degrees of freedom. Now, from the Central Limit Theorem, we have that

$$\frac{p\|\mathbf{Z}\|^2 - p}{\sqrt{2p}} = \frac{\sqrt{p}}{\sqrt{2}}\left(\|\mathbf{Z}\|^2 - 1\right)$$

converges weakly to a standard Gaussian random variable as $p \to \infty$. It therefore follows that $\|\mathbf{Z}\|^2 = 1 + O_{\mathrm{P}}(p^{-1/2})$ as $p \to \infty$ so that $\mathbf{Z}$ is almost surely a point on the unit hypersphere as $p \to \infty$. Note that this happens despite the fact that the origin is the point with highest density. As a consequence, Gaussian vectors and directional vectors are closely related in the high-dimensional setup.

Providing high-dimensional models for directional data is a non-trivial task. This is mainly due to the fact that the surface area of the unit hypersphere

$$\omega_p = \frac{2\pi^{p/2}}{\Gamma(p/2)} \to 0$$

as the dimension p increases,[1] combined with measure concentration effects. For instance, recall formula (2.22), the density of $\mathbf{X}'\boldsymbol{\mu}$ when $\mathbf{X}$ is rotationally symmetric around $\boldsymbol{\mu} \in \mathcal{S}^{p-1}$. One readily sees that the factor $(1-t^2)^{(p-3)/2}$ tends to zero except for $t = 0$, yielding a concentration around the value $\mathbf{X}'\boldsymbol{\mu} = 0$. Thus, in high dimensions, all classical models of Section 2.3.1 tend to behave like girdle distributions, i.e., they are concentrated around the equator with respect to $\boldsymbol{\mu}$. One way to overcome this feature is to introduce a dimension-related concentration, as done below.

High-dimensional distributions on $\mathcal{S}^{p-1}$ were proposed and studied in Dryden (2005), who used a relation with Wiener measures. Letting $\mathcal{C} := \{h \in C[0,1] : h(0) = 0\}$ be the set of continuous paths on $[0,1]$ starting at zero, define

$$Q_p(\mathbf{x}, k/p) := \sum_{i=1}^{k} x_i, \tag{7.2}$$

[1]Note that the mapping $p \mapsto \omega_p$ is not monotone and has a peak at $p = 7$.

where $\mathbf{x} = (x_1, \ldots, x_p)' \in \mathcal{S}^{p-1}$. The path $Q_p(\mathbf{x}, \cdot)$ in (7.2) is well defined on $\mathcal{C}$ since $Q_p(\mathbf{x}, 0) = 0$. It follows from Cutland & Ng (1993) that when $\mathbf{X}$ is uniformly distributed on $\mathcal{S}^{p-1}$, $Q_p(\mathbf{X}, k/p)$ tends to a Wiener process on $\mathcal{C}$ as $p \to \infty$. This relation between the uniform measure on $\mathcal{S}^{\infty}$ and the Wiener measure can be exploited to provide probability measures on $\mathcal{S}^{\infty}$. As shown in Dryden (2005) if $\mathbf{X}$ follows a FvML or a Watson distribution with density

$$\mathbf{x} \mapsto \frac{\left(\frac{p^{1/2}\kappa}{2}\right)^{p/2-1}}{2\pi^{p/2} I_{p/2-1}(p^{1/2}\kappa)} \exp(p^{1/2}\kappa \mathbf{x}'\boldsymbol{\mu})$$

or

$$\mathbf{x} \mapsto \frac{\Gamma(p/2)}{2\pi^{p/2} M\left(\frac{1}{2}, \frac{p}{2}, p\kappa\right)} \exp(p\kappa(\mathbf{x}'\boldsymbol{\mu})^2)$$

respectively and if $\mathbf{P}_h$ is a $p \times h$ matrix with h orthogonal columns, we have that $p^{1/2}\mathbf{P}_h'\mathbf{X}$ is asymptotically normal as $p \to \infty$. We refer the reader to Dryden (2005) for details on the properties of high-dimensional FvML and Watson distributions.

7.3 Testing uniformity in the high-dimensional case

Before entering into problem-specific details, we describe a quite general high-dimensional testing setup. Consider an asymptotic (as $n \to \infty$) fixed-p test $\phi^{(n)}$ that rejects some null hypothesis $\mathcal{H}_0$ at asymptotic level α whenever some statistic $Q^{(n)} > \chi^2_{\ell(p);1-\alpha}$, where the mapping $p \mapsto \ell(p)$ is monotone increasing. The Central Limit Theorem teaches us that if some random variable Y_d is chi-squared with d degrees of freedom, then $(Y_d - d)/\sqrt{2d}$ converges weakly to the standard normal distribution as $d \to \infty$. Thus it may be *expected* that the test $\phi_p^{(n)}$ rejecting $\mathcal{H}_0$ whenever $Q^{\mathrm{St}} = \frac{Q^{(n)} - \ell(p)}{\sqrt{2\ell(p)}} > z_{1-\alpha}$, the α-upper quantile of the standard normal distribution, has asymptotic level α under $\mathcal{H}_0$ when both n and p converge to infinity. Of course, establishing that the sequence of tests $\phi_p^{(n)}$ — hence also the sequence of tests $\phi^{(n)}$ — is valid in the high-dimensional setup, that is, has asymptotic level α under the null when both n and p tend to infinity, requires a formal proof. All the more so as the result does not always hold true: an (n, p)-limit cannot always correspond to the limit obtained by first letting n go to infinity, p fixed, and then let p grow to infinity. Some test statistics indeed need to be appropriately corrected to obtain valid (n, p)-asymptotic results, while some others do not. We refer the interested reader to Ledoit & Wolf (2002) where precisely two such situations are addressed. The test statistics that do not need to be corrected can be called *high-dimensional (HD-)robust* following the Paindaveine & Verdebout (2016) terminology. The test statistic $Q^{(n)}$ is thus HD-robust if, under the null hypothesis, $\frac{Q^{(n)}-\ell(p_n)}{\sqrt{2\ell(p_n)}}$ converges weakly to a standard Gaussian random variable as n and $p = p_n$ tend to infinity.

Now, consider iid random vectors $\mathbf{X}_1, \ldots, \mathbf{X}_n$ taking values on $\mathcal{S}^{p-1}$. We described in Section 6.2 the classical Rayleigh test of uniformity, rejecting $\mathcal{H}_0^{\text{unif}}$ for large values of

$$R_n = \frac{p}{n} \sum_{i,j=1}^{n} \mathbf{X}_i' \mathbf{X}_j$$

which converges to a χ_p^2 distribution as $n \to \infty$. Here we discuss how this statistic behaves when the dimension p becomes large, too. To this end, first reexpress it as

$$R_n = \frac{p}{n}\left(n + \sum_{1 \le i \ne j \le n} \mathbf{X}_i' \mathbf{X}_j\right) = p + \frac{2p}{n} \sum_{1 \le i < j \le n} \mathbf{X}_i' \mathbf{X}_j,$$

and then consider the standardized statistic

$$R_n^{\text{St}} = \frac{R_n - p}{\sqrt{2p}} = \frac{\sqrt{2p}}{n} \sum_{1 \le i < j \le n} \mathbf{X}_i' \mathbf{X}_j. \tag{7.3}$$

It was shown in Paindaveine & Verdebout (2016) that, under minimal assumptions, R_n^{St} does converge to a standard normal distribution as $n, p \to \infty$. More precisely, using a central limit theorem for martingale differences, they show that R_n^{St} is asymptotically normal as $\min(n, p) \to \infty$. As a consequence, the Rayleigh test is HD-robust. This convergence can be appreciated from a consideration of the histograms in Figure 7.1, where the behavior under $\mathcal{H}_0^{\text{unif}}$ is analyzed by means of Monte Carlo simulations for various combinations of n and p. Regarding the power of the Rayleigh test in high-dimensional settings, Cutting et al. (2017*a*) established that, against FvML alternatives with concentration $\kappa_n \sim p_n^{3/4}/\sqrt{n}$, the Rayleigh test has non-trivial powers (neither α nor 1) as $n \to \infty$ and $p_n \to \infty$. It furthermore enjoys the property of being locally and (n,p)-asymptotically most powerful invariant.

A different approach to test uniformity, based on the behavior of the random cosines $\rho_{ij} := \mathbf{X}_i' \mathbf{X}_j$ $(i \neq j)$, was considered by Cai & Jiang (2012) and Cai, Fan & Jiang (2013). The ρ_{ij}'s play an important role in the study of the coherence of random matrices. The coherence of a random matrix is defined as the largest magnitude of the off-diagonal elements of the sample correlation matrix generated

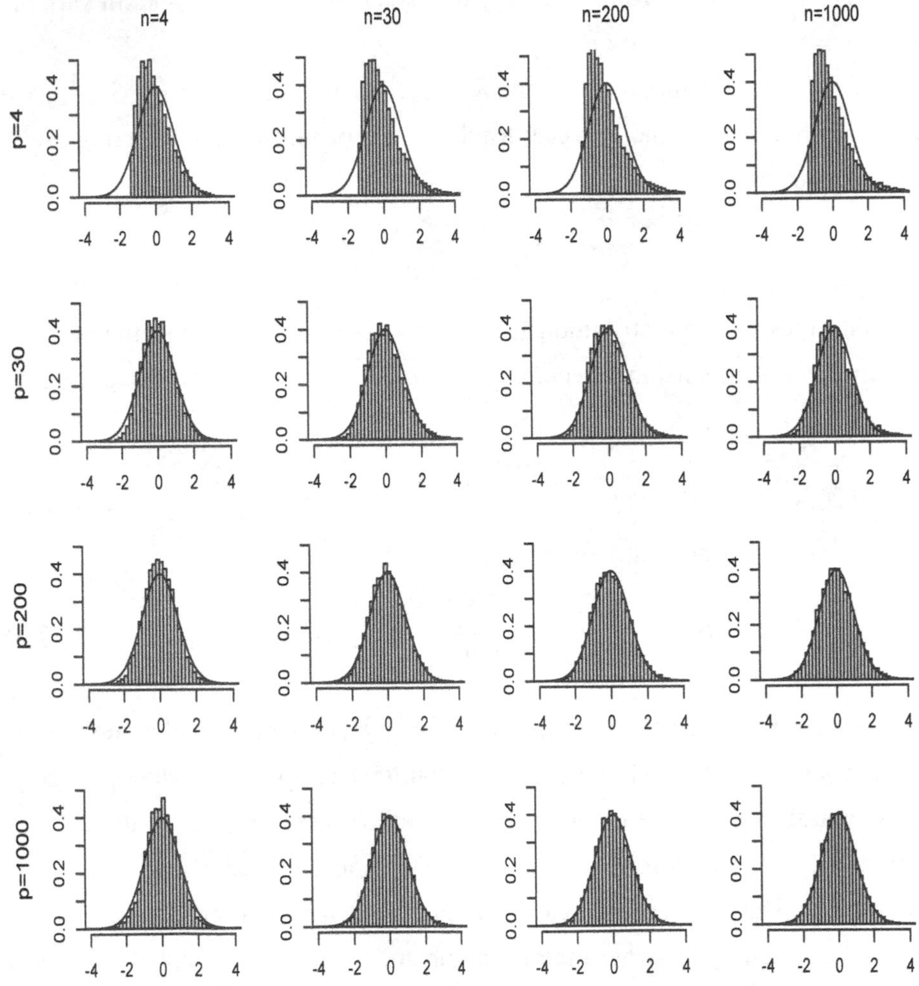

Figure 7.1: Histograms, for the specified values of n and p, of the standardized Rayleigh test statistic R^{St} calculated using $M = 10{,}000$ random samples of size n from the uniform distribution on $\mathcal{S}^{p-1}$. The standard Gaussian density (black) appears superimposed in each frame.

from a random matrix. The quantity of interest is therefore

$$\ell_n := \max_{1 \leq i < j \leq n} |\rho_{ij}|.$$

Contrary to the high-dimensional Rayleigh test, the limiting distribution of ℓ_n depends crucially on how $p = p_n$ goes to infinity as a function of n. Three distinct regimes are considered:

(i) *The sub-exponential regime*: the positive sequence p_n is such that $\log(p_n)/n \to 0$ as $n \to \infty$.

(ii) *The exponential regime*: the positive sequence p_n is such that $\log(p_n)/n \to \beta \in (0, +\infty)$ as $n \to \infty$.

(iii) *The super-exponential regime*: the positive sequence p_n is such that $\log(p_n)/n \to +\infty$ as $n \to \infty$.

In the sub-exponential regime, Cai & Jiang (2012) showed that $\ell_n \to 0$ in probability as $n \to \infty$ and, letting $T_n := \log(1 - \ell_n^2)$, that the statistic

$$nT_n + 4\log p_n - \log\log p_n$$

converges weakly to an extreme value distribution with distribution function

$$z \mapsto 1 - \exp\left(-\frac{1}{\sqrt{8\pi}} \exp\left(\frac{z}{2}\right)\right).$$

The speed of convergence of ℓ_n to zero follows from the fact that $\sqrt{\frac{n}{\log(p_n)}}\ell_n \to 2$ in probability as $n \to \infty$. In the exponential case, ℓ_n converges to $\sqrt{1 - \exp(-4\beta)}$ in probability as $n \to \infty$ and $nT_n + 4\log p_n - \log\log p_n$ converges weakly to a random variable with distribution function

$$z \mapsto 1 - \exp\left(-\sqrt{\frac{\beta}{2\pi(1 - \exp(-4\beta))}} \exp\left(\frac{z + 8\beta}{2}\right)\right)$$

as $n \to \infty$. Since $\sqrt{\frac{\beta}{2\pi(1-\exp(-4\beta))}}$ tends to $\frac{1}{\sqrt{8\pi}}$ as $\beta \to 0$, the result is consistent with that of the sub-exponential regime. Finally, in the super-exponential case, $\ell_n \to 1$ in probability as $n \to \infty$ and $nT_n + \frac{4n}{n-2}\log p_n - \log n$ converges weakly to a random variable with distribution function

$$z \mapsto 1 - \exp\left(-\frac{1}{\sqrt{2\pi}} \exp\left(\frac{z}{2}\right)\right).$$

All these results can be used to derive tests of uniformity. For instance, the null hypothesis can be rejected if ℓ_n is too large. The critical values of the different tests can be obtained from the asymptotic results for the various regimes provided above.

7.4 Location tests in the high-dimensional case

Let $\mathbf{X}_1, \ldots, \mathbf{X}_n$ be a random sample from a rotationally symmetric distribution around $\boldsymbol{\mu} \in \mathcal{S}^{p-1}$ and consider the location testing problem $\mathcal{H}_0 : \boldsymbol{\mu} = \boldsymbol{\mu}_0$ against the alternative $\mathcal{H}_1 : \boldsymbol{\mu} \neq \boldsymbol{\mu}_0$, where $\boldsymbol{\mu}_0 \in \mathcal{S}^{p-1}$ is fixed. The underlying distribution of the observations is unspecified. Letting $\bar{\mathbf{X}} = \frac{1}{n}\sum_{i=1}^{n} \mathbf{X}_i$, the classical test for this location problem rejects the null hypothesis for large values of the Watson statistic (Watson 1983)

$$W_n := \frac{n(p-1)\bar{\mathbf{X}}'(\mathbf{I}_p - \boldsymbol{\mu}_0\boldsymbol{\mu}_0')\bar{\mathbf{X}}}{1 - \frac{1}{n}\sum_{i=1}^{n}(\mathbf{X}_i'\boldsymbol{\mu}_0)^2}. \tag{7.4}$$

Under very mild assumptions on the underlying distribution, the fixed-p asymptotic null distribution of W_n is χ^2_{p-1}. The resulting test ϕ_n^W therefore rejects the null hypothesis, at asymptotic level α, whenever $W_n > \chi^2_{p-1;1-\alpha}$. As well as achieving asymptotic level α under virtually any rotationally symmetric distribution, ϕ_n^W is optimal — more precisely, locally and asymptotically maximin (see Chapter 5)— when the underlying distribution is FvML. We refer the interested reader to Paindaveine & Verdebout (2015) for details.

The high-dimensional version of ϕ_n^W was investigated in Ley et al. (2015). Consider a triangular array of observations $\mathbf{X}_{ni}$, $i = 1, \ldots, n$, $n = 1, 2, \ldots$, where $\mathbf{X}_{ni}$ takes values on $\mathcal{S}^{p_n-1}$, p_n tends to infinity with n and define

$$u_{ni} := \sqrt{1 - (\mathbf{X}_{ni}'\boldsymbol{\mu}_0)^2}.$$

Assume the following extremely mild conditions:

(i) for any n, $\mathbf{X}_{n1}, \mathbf{X}_{n2}, \ldots, \mathbf{X}_{nn}$ are mutually independent and share a common rotationally symmetric distribution on $\mathcal{S}^{p_n-1}$ with location $\boldsymbol{\mu}_0$;

(ii) $\mathrm{E}[u_{n1}^2] > 0$, which only excludes the degenerate case when the $\mathbf{X}_{ni}$'s are located at $\boldsymbol{\mu}_0$ almost surely;

(iii) $\mathrm{E}[u_{n1}^4]/(\mathrm{E}[u_{n1}^2])^2 = o(n)$ as $n \to \infty$.

The latter condition deserves some comments. A sufficient condition for (iii) is that $\sqrt{n}\mathrm{E}[u_{n1}^2] \to \infty$ as $n \to \infty$. In other words, if (iii) does not hold, then there exists some constant $C > 0$ such that

$$\mathrm{E}[(\mathbf{X}_{n1}'\boldsymbol{\mu}_0)^2] \geq 1 - \frac{C}{\sqrt{n}}$$

for infinitely many n, meaning that the distribution of each $\mathbf{X}_{ni}$ concentrates in the direction $\boldsymbol{\mu}_0$ in the expanding Euclidean space $\mathbb{R}^{p_n}$. Condition (iii) rules out these pathological settings. It is satisfied, for instance, by the FvML distribution for which Ley et al. (2015) established that $\mathrm{E}[u_{n1}^4]/(\mathrm{E}[u_{n1}^2])^2 \leq 3$.

Under conditions (i)–(iii), and without any particular requirements on how p_n tends to infinity as a function of n, we have that

$$\tilde{W}_n := \frac{W_n - (p_n - 1)}{\sqrt{2(p_n - 1)}} \xrightarrow{\mathcal{D}} \mathcal{N}(0, 1) \tag{7.5}$$

as $n \to \infty$. We therefore say that (7.5) is an (n, p)-universal result, which further shows that the Watson test is HD-robust in the sense of the previous section. The speed of convergence is illustrated in Figure 7.2 where the behavior of the standardized Watson statistic $\tilde{W}_n$ under $\mathcal{H}_0$ is illustrated using the results from a Monte Carlo study for various combinations of n and p.

The high-dimensional location testing problem was also addressed in Paindaveine & Verdebout (2016) who considered the sign statistic

$$\tilde{Q}_n := \frac{\sqrt{2(p_n - 1)}}{n} \sum_{1 \leq i < j \leq n} \mathbf{S}_{\boldsymbol{\mu}_0}(\mathbf{X}_{ni})' \mathbf{S}_{\boldsymbol{\mu}_0}(\mathbf{X}_{nj}).$$

They proved that $\tilde{Q}_n$ is also (n, p)-universally asymptotically standard normal under mild conditions. This raises the question as to when $\tilde{W}_n$ and $\tilde{Q}_n$ are asymptotically equivalent in probability under $\mathcal{H}_0$. The answer is quite intuitive: it happens in the rare cases when the u_{ni}'s become asymptotically constant, in the sense that $\mathrm{Var}[u_{ni}]/(\mathrm{E}[u_{ni}])^2 \to 0$ as $n \to \infty$. We refer to Ley et al. (2015) for details.

7.5 Concentration tests in the high-dimensional case

Let $\mathbf{X}_1, \ldots, \mathbf{X}_n$ be an iid sequence of FvML random vectors with common location $\boldsymbol{\mu}$ and concentration κ. We consider the problem of testing $\mathcal{H}_0 : \kappa = \kappa_0$ against $\mathcal{H}_1 : \kappa \neq \kappa_0$, where $\kappa_0 > 0$ is a fixed value. It is well known that

$$e_1 := \mathrm{E}[\mathbf{X}_i' \boldsymbol{\mu}] = \frac{\mathcal{I}_{p/2}(\kappa)}{\mathcal{I}_{p/2-1}(\kappa)} = A_p(\kappa).$$

The function $A_p(\kappa)$ takes values in $[0, 1]$ and plays a crucial role in maximum likelihood estimation for FvML distributions; see Section 4.4 of Chapter 4. Since it is a one-to-one mapping, concentration may equivalently be measured through e_1, and one may rephrase the null hypothesis $\mathcal{H}_0 : \kappa = \kappa_0$ as $\mathcal{H}_0 : e_1 = e_{10}$,

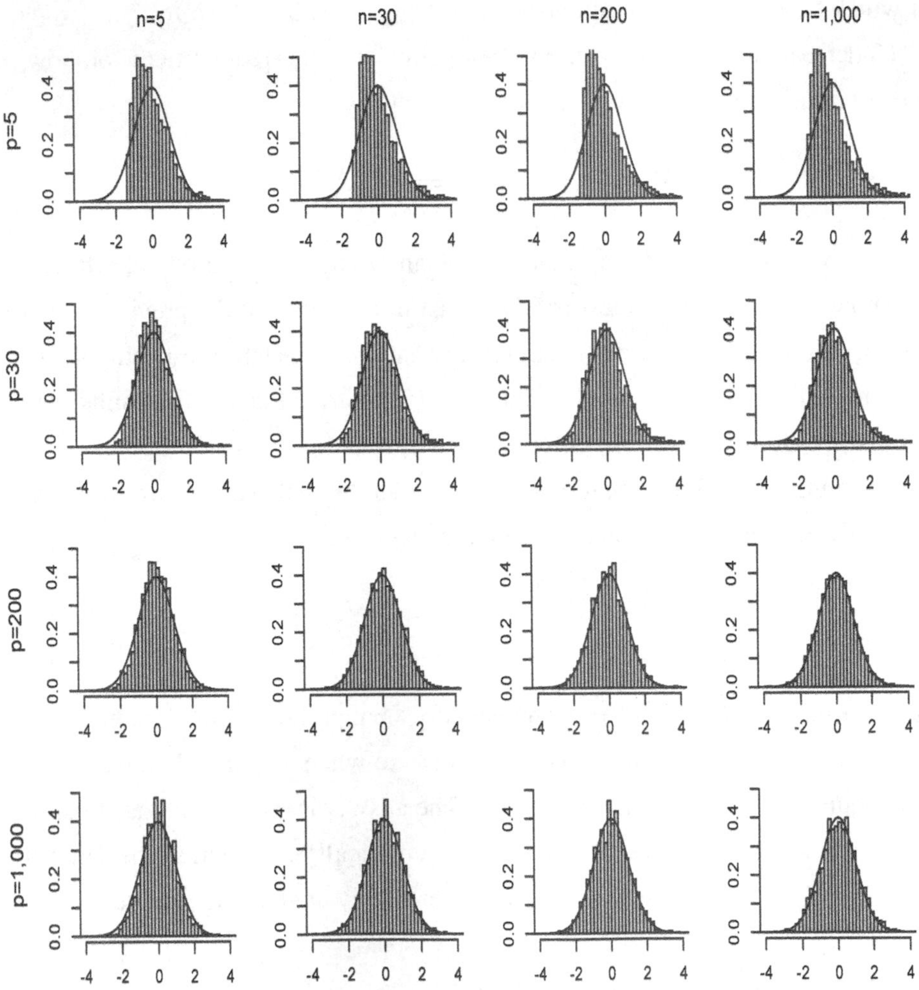

Figure 7.2: Histograms, for the specified values of n and p, of the standardized Watson statistic $\tilde{W}_n$ calculated using $M = 2{,}500$ random samples of size n from the p-dimensional FvML distribution with concentration parameter $\kappa = 2$. The standard Gaussian density (black) appears superimposed in each frame.

with $e_{10} := A_p(\kappa_0)$. In Section 5.3.5 we saw that the most efficient test for the concentration problem rejects the null hypothesis at asymptotic level α whenever

$$Q_{\kappa_0}^{(n)} := \frac{n(\|\bar{\mathbf{X}}\| - e_{10})^2}{1 - \frac{p-1}{\kappa_0} e_{10} - e_{10}^2} > \chi^2_{1;1-\alpha}.$$

This is a slight modification of the test statistic (5.26), obtained by replacing the general root-n consistent estimator $\hat{\boldsymbol{\mu}}^{(n)}$ by the sample mean $\bar{\mathbf{X}}/\|\bar{\mathbf{X}}\|$. Unlike the Watson test for the location problem, the concentration test based on $Q_{\kappa_0}^{(n)}$ is not HD-robust. This was established in Cutting et al. (2017*b*). The non-robustness is strongly linked to the concentration effects in high dimensions mentioned in Section 7.2, and is further strengthened by the fact that $e_{10} = \frac{I_{p/2}(\kappa_0)}{I_{p/2-1}(\kappa_0)}$ converges to zero as $p \to \infty$, a result that can be deduced from Section 4.4. This implies that the one-to-one correspondence between a test for e_1 and a test for κ does not hold in the high-dimensional case. Therefore, rather than considering fixed values κ_0, it is sensible to tackle the concentration testing problem $\mathcal{H}_0 : e_{n1} = e_{10}$ under sequences of FvML distributions indexed by a sequence of concentrations κ_n such that $\kappa_n \to \infty$ as $n \to \infty$. In other words, we work with triangular arrays of observations $\mathbf{X}_{ni}$, $i = 1, \ldots, n$, $n = 1, 2, \ldots$ such that, for any n, $\mathbf{X}_{n1}, \mathbf{X}_{n2}, \ldots, \mathbf{X}_{nn}$ are iid FvML with location $\boldsymbol{\mu}_n$ and concentration κ_n.

We are thus here dealing with a setting where the sample size n, the dimension p_n and the concentration κ_n all tend to infinity. This raises the natural question of the speed of convergence of κ_n compared to that of p_n. As the following result indicates, the convergence regime plays a fundamental role.

Proposition 7.5.1 (Cutting et al. 2017*b*) *Let $(p_n)_n$ be a sequence of positive integers converging to ∞, $(\boldsymbol{\mu}_n)_n$ be an arbitrary sequence such that $\boldsymbol{\mu}_n \in \mathcal{S}^{p_n-1}$ for any n, and $(\kappa_n)_n$ be a sequence in $(0, \infty)$. Write $e_{n1} := \mathrm{E}[\mathbf{X}_{n1}'\boldsymbol{\mu}_n]$ and $\tilde{e}_{n2} := \mathrm{Var}[\mathbf{X}_{n1}'\boldsymbol{\mu}_n]$. We have the following results (all convergences are as $n \to \infty$) :*

(i) $\kappa_n/p_n \to 0 \Longleftrightarrow e_{n1} \to 0$;

(ii) $\kappa_n/p_n \to c \in (0, \infty) \Longleftrightarrow e_{n1} \to g_1(c)$, *where* $g_1 : (0, \infty) \to (0, 1) : z \mapsto z/\left(\frac{1}{2} + (z^2 + \frac{1}{4})^{1/2}\right)$ *is one-to-one;*

(iii) $\kappa_n/p_n \to \infty \Longleftrightarrow e_{n1} \to 1$.

In cases (i) and (iii), $\tilde{e}_{n2} \to 0$, whereas in case (ii), $\tilde{e}_{n2} \to g_2(c)$, for some function $g_2 : (0, \infty) \to (0, 1)$.

These convergence results are very intuitive. In setting (i), the concentration grows too slowly compared to the dimension, hence the observations cluster around the equator with respect to $\boldsymbol{\mu} = \lim_{n\to\infty} \boldsymbol{\mu}_n$. Setting (iii) is exactly the opposite, and all observations tend to coincide with $\boldsymbol{\mu}$. In both settings, it is natural that the variance vanishes. The only non-trivial case thus is setting (ii).

A direct consequence of Proposition 7.5.1 is that the values $e_{10} \in (0,1)$ can only be tested when $\kappa_n \sim p_n$. In the other two cases, the only values of e_{10} which are admissible are either zero or one. While the null hypothesis $\mathcal{H}_0 : e_{n1} = 0$ coincides with the null hypothesis of uniformity, the null $\mathcal{H}_0 : e_{n1} = 1$ is extremely pathological since it implies that all the observations have to coincide. In such a setting, a better and simpler test would reject the null hypothesis when there exists one $\mathbf{X}_{ni}$ such that $\mathbf{X}_{ni} \neq \mathbf{X}_{nj}$ for some $j \neq i$. The only meaningful concentration problem thus corresponds to the second situation in Proposition 7.5.1. For this problem, Cutting et al. (2017*b*) proposed the test $\phi^{(n)}_{\text{CPV}}$ that rejects $\mathcal{H}_0 : e_{n1} = e_{10}$ at asymptotic level α whenever

$$|Q^{(n)}_{\text{CPV}}| > z_{1-\alpha/2},$$

where

$$Q^{(n)}_{\text{CPV}} := \frac{\sqrt{p_n}\big(n\|\bar{\mathbf{X}}_n\|^2 - 1 - (n-1)e_{10}^2\big)}{\sqrt{2}\Big(p_n\big(1 - \frac{e_{10}}{c_0} - e_{10}^2\big)^2 + 2np_ne_{10}^2\big(1 - \frac{e_{10}}{c_0} - e_{10}^2\big) + \big(\frac{e_{10}}{c_0}\big)^2\Big)^{1/2}},$$

with

$$c_0 = g_1^{-1}(e_{10}).$$

This test is valid in the high-dimensional setting, and $Q^{(n)}_{\text{CPV}}$ can be perceived as a high-dimensional modification of $Q^{(n)}_{\kappa_0}$.

7.6 Principal nested spheres

Principal Component Analysis is one of the most important tools in multivariate analysis. The main objective of PCA is dimension reduction: the original dataset may involve a considerable number of correlated variables that can often be summarized using far fewer uncorrelated variables called the principal components.

The analysis of Principal Nested Spheres (PNS) is a decomposition method introduced by Jung et al. (2012). It yields a sequence of submanifolds $\mathcal{A}_1, \mathcal{A}_2, \ldots, \mathcal{A}_{p-2}$ of $\mathcal{S}^{p-1}$ such that

$$\mathcal{A}_1 \subset \mathcal{A}_2 \subset \cdots \subset \mathcal{A}_{p-2} \subset \mathcal{S}^{p-1}.$$

The terminology follows from the fact that these submanifolds will be identified with unit spheres in lower dimensions, as we shall now explain. The geodesic distance d_g on the unit sphere is defined as

$$d_g(\mathbf{x}, \mathbf{y}) := \arccos(\mathbf{x}'\mathbf{y}), \quad \mathbf{x}, \mathbf{y} \in \mathcal{S}^{p-1},$$

which is the length of the shortest great circle segment (geodesic) joining $\mathbf{x}$ and $\mathbf{y}$. Note that the shortest path is unique unless $\mathbf{x}'\mathbf{y} = -1$. A subsphere A_{p-2} of $\mathcal{S}^{p-1}$ is defined in terms of an axis $\mathbf{v}$ and a distance $r \in (0, \pi/2]$ as

$$A_{p-2}(\mathbf{v}, r) := \{\mathbf{x} \in \mathcal{S}^{p-1} : d_g(\mathbf{x}, \mathbf{v}) = r\}.$$

The sphere A_{p-2} is nothing other than the intersection between $\mathcal{S}^{p-1}$ and a hyperplane of the form $\{\mathbf{z} \in \mathbb{R}^p | \mathbf{v}'\mathbf{z} = \cos(r)\}$; it is therefore a $(p-2)$-dimensional sphere nested within $\mathcal{S}^{p-1}$. See Figure 7.3 for an illustration when $p = 3$. It is, in general, not a unit sphere. The subsphere A_{p-2} can be mapped onto the unit hypersphere $\mathcal{S}^{p-2}$ via a simple invertible mapping f_1 (see Jung et al. 2012 for its expression), hence it can be identified with $\mathcal{S}^{p-2}$ and treated as a unit sphere. The next subsphere A_{p-3} can then be built from $\mathcal{S}^{p-2}$ (given a new axis $\mathbf{v}^* \in \mathcal{S}^{p-2}$ and a new distance $r^* \in (0, \pi/2]$) and identified with a submanifold of $\mathcal{S}^{p-1}$ via $f_1^{-1}(A_{p-3}) \subset A_{p-2}$. Continuing this process, we can define the $(p-k)$-dimensional nested sphere of $\mathcal{S}^{p-1}$ as

$$A_{p-k} = \begin{cases} f_1^{-1} \circ \ldots \circ f_{k-2}^{-1}(A_{p-k}) & \text{if } k = 3, \ldots, p-1 \\ A_{p-2} & \text{if } k = 2, \end{cases}$$

where the f_j's are the consecutive invertible mappings linking subspheres A_{p-j-1} with unit spheres $\mathcal{S}^{p-j-1}$.

Now, let $\mathbf{X}_1, \ldots, \mathbf{X}_n$ be a sample taking values on $\mathcal{S}^{p-1}$. The residual of a point $\mathbf{x} \in \mathcal{S}^{p-1}$ for the subsphere $A_{p-2}(\mathbf{v}_1, r_1)$ is the (signed) length of the geodesic that joins $\mathbf{x}$ to $A_{p-2}(\mathbf{v}_1, r_1)$ given by $d_g(\mathbf{x}, \mathbf{v}_1) - r_1$. The best fitting least squares subsphere is then based on

$$(\hat{\mathbf{v}}_1, \hat{r}_1) := \text{argmin}_{(\mathbf{v}, r) \in (\mathcal{S}^{p-1} \times (0, \pi/2])} \sum_{i=1}^{n} (d_g(\mathbf{X}_i, \mathbf{v}) - r)^2. \tag{7.6}$$

The choice of the objective function in (7.6) is arbitrary and the robustness of the estimator could probably be improved by choosing other types of objective functions. The $(p-2)$-dimensional empirical nested sphere $\hat{A}_{p-2}$ is therefore

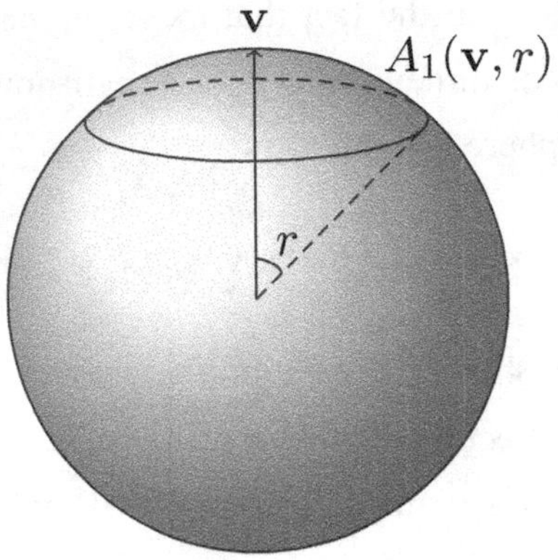

Figure 7.3: Illustration of a nested sphere.

$A_{p-2}(\hat{\mathbf{v}}_1, \hat{r}_1)$. The data are then projected onto this nested sphere along the minimal geodesic joining each observation to $\hat{\mathcal{A}}_{p-2}$, and the same process as in (7.6) is applied to these projected data to find the second best fitting sphere $A_{p-3}(\hat{\mathbf{v}}_2, \hat{r}_2)$ and hence $\hat{\mathcal{A}}_{p-3} \subset \hat{\mathcal{A}}_{p-2} \subset \mathcal{S}^{p-1}$.

Principal nested spheres have been used for various applications such as the analysis of the migration paths of elephants, sea sand grains and human movements, to cite but these.

7.7 Further reading

We started this chapter by indicating how the distributions on $\mathcal{S}^{p-1}$ need to be adapted to the setting of high-dimensional spheres. In particular, Section 7.2 describes the high-dimensional FvML and Watson distributions. Further well-known distributions have been analyzed in high-dimensional settings by Dryden (2005), including the Bingham or Fisher-Bingham models.

The high-dimensional Watson test of Section 7.4 can be used in the solution of non-directional problems. Indeed, as shown in Ley et al. (2015), it happens to be as well a valid procedure for testing if the covariance matrix of high-dimensional random vectors on $\mathbb{R}^p$ has a spiked form as in (7.1).

Bibliography

Abe, T. & Ley, C. (2017), 'A tractable, parsimonious and highly flexible model for cylindrical data, with applications'. *EcoSta*, forthcoming.

Abe, T. & Pewsey, A. (2011*a*), 'Sine-skewed circular distributions', *Stat. Papers* **52**, 683–707.

Abe, T. & Pewsey, A. (2011*b*), 'Symmetric circular models through duplication and cosine perturbation', *Comp. Stat. Data Anal.* **55**, 3271–3282.

Abe, T., Pewsey, A. & Shimizu, K. (2009), 'On Papakonstantinou's extension of the cardioid distribution', *Stat. Probab. Lett.* **79**, 2138–2147.

Abe, T., Pewsey, A. & Shimizu, K. (2013), 'Extending circular distributions through transformation of argument', *Ann. Inst. Statist. Math.* **65**, 833–858.

Abe, T., Shimizu, K. & Pewsey, A. (2010), 'Symmetric unimodal models for directional data motivated by inverse stereographic projection', *J. Japan Statist. Soc.* **40**, 45–61.

Abramowitz, M. & Stegun, I. A. (1965), *Handbook of Mathematical Functions*, Dover, New York.

Agostinelli, C. & Romanazzi, M. (2013), 'Nonparametric analysis of directional data based on data depth', *Environ. Ecol. Stat.* **20**, 253–270.

Ajne, B. (1968), 'A simple test for uniformity of a circular distribution', *Biometrika* **55**, 343–354.

Amos, D. E. (1974), 'Computation of modified Bessel functions and their ratios', *Math. Comput.* **28**, 239–251.

Arnold, K. J. (1941), On Spherical Probability Distributions, PhD thesis, Massachusetts Institute of Technology.

Azzalini, A. & Capitanio, A. (2014), *The Skew-Normal and Related Families*, Cambridge, IMS Monographs, Cambridge University Press.

Baayen, C. & Klugkist, I. G. (2014), 'Evaluating order-constrained hypotheses for circular data from a between-within subjects design', *Psychol. Methods* **19**, 398–408.

Baayen, C., Klugkist, I. G. & Mechsner, F. (2012), 'A test of order constrained hypotheses for circular data with applications to human movement science', *J. Motor. Behav.* **44**, 351–363.

Bai, Z. D., Rao, C. R. & Zhao, L. C. (1988), 'Kernel estimators of density function of directional data', *J. Multivariate Anal.* **27**, 24–39.

Bai, Z., Jiang, D., Yao, J.-F. & Zheng, S. (2009), 'Corrections to LRT on large-dimensional covariance matrix by RMT', *Ann. Statist.* **37**, 3822–3840.

Baldi, P., Kerkyacharian, G., Marinucci, D. & Picard, D. (2009), 'Adaptive density estimation for directional data using needlets', *Ann. Statist.* **37**, 3362–3395.

Banerjee, A., Dhillon, I., Ghosh, J. & Sra, S. (2005), 'Clustering on the unit hypersphere using von Mises-Fisher distributions', *J. Mach. Learn. Res.* **6**, 1345–1382.

Baricz, A. (2014), 'Remarks on a parameter estimation for von Mises-Fisher distributions', *Comput. Stat.* **29**, 891–894.

Baringhaus, L. (1991), 'Testing for spherical symmetry of a multivariate distribution', *Ann. Statist.* **19**, 899–917.

Barlow, R. E., Bartholomew, R. J., Bremner, J. M. & Brunk, H. D. (1972), *Statistical Inference under Order Restrictions*, Wiley, New York.

Barragán, S., Fernández, M., Rueda, C. & Peddada, S. (2013), 'isocir: An R package for constrained inference using isotonic regression for circular data, with an application to cell biology', *J. Stat. Softw.* **54**, 1–17.

Barragán, S., Rueda, C., Fernández, M. & Peddada, S. (2015), 'Determination of temporal order among the components of an oscillatory system', *PLoS ONE* **10**, e0124842.

Barros, A. M. G., Pereira, J. M. C. & Lund, U. J. (2012), 'Identifying geographical patterns of wildfire orientation: A watershed-based analysis', *Forest. Ecol. Manag.* **264**, 98–107.

Batschelet, E. (1965), *Statistical Methods for the Analysis of Problems in Animal Orientation and Certain Biological Rhythms*, American Institute of Biological Sciences, Washington, DC.

Batschelet, E. (1981), *Circular Statistics in Biology*, Academic Press, London.

Begun, J. M., Hall, W., Huang, W.-M. & Wellner, J. A. (1983), 'Information and asymptotic efficiency in parametric-nonparametric models', *Ann. Statist.* **11**, 432–452.

Beran, R. (1968), 'Testing for uniformity on a compact homogeneous space', *J. Appl. Probab.* **5**, 177–195.

Beran, R. (1969), 'Asymptotic theory of a class of tests for uniformity of a circular distribution', *Ann. Math. Stat.* **40**, 1196–1206.

Beran, R. (1979*a*), 'Exponential models for directional data', *Ann. Statist.* **7**, 1162–1178.

Beran, R. (1979*b*), 'Testing for ellipsoidal symmetry of a multivariate density', *Ann. Statist.* **7**, 150–162.

Bernoulli, D. (1735), Quelle est la cause physique de l'inclination des plans des orbites des planètes?, *in* 'Recueil des pièces qui ont remporté le prix de l'Académie Royale des Sciences de Paris 1734, 93–122', Académie Royale des Sciences de Paris, Paris. Reprinted in *Daniel Bernoulli, Werke, Vol. 3, 226–303, Birkhäuser, Basel (1982)*.

Berthet, Q. & Rigollet, P. (2013), 'Optimal detection of sparse principal components in high dimension', *Ann. Statist.* **41**, 1780–1815.

Bhattacharya, S. & SenGupta, A. (2009), 'Bayesian analysis of semiparametric linear-circular models', *J. Agr. Biol. Envir. St.* **14**, 33–65.

Bickel, P. J. & Levina, E. (2004), 'Some theory for Fisher's linear discriminant function,'naive Bayes', and some alternatives when there are many more variables than observations', *Bernoulli* **10**, 989–1010.

Bickel, P., Klaassen, C., Ritov, Y. & Wellner, J. (1993), *Efficient and Adaptive Estimation for Semiparametric Models*, Springer Verlag, New York.

Bingham, C. (1964), Distributions on the Sphere and on the Projective Plane, PhD thesis, Yale University.

Bingham, C. (1974), 'An antipodally symmetric distribution on the sphere', *Ann. Statist.* **2**, 1201–1225.

Bingham, M. S. & Mardia, K. V. (1975), Maximum likelihood characterization of the von Mises distribution, *in* G. P. Patil, S. Kotz & J. K. Ord, eds, 'Statistical Distributions for Scientific Work', Vol. 3, Reidel, Dordrecht and Boston, pp. 387–398.

Birke, M. & Dette, H. (2005), 'A note on testing the covariance matrix for large dimension', *Stat. Probab. Lett.* **74**, 281–289.

Boente, G., Rodriguez, D. & González-Manteiga, W. (2014), 'Goodness-of-fit test for directional data', *Scand. J. Statist.* **41**, 259–275.

Boulerice, B. & Ducharme, G. R. (1997), 'Smooth tests of goodness-of-fit for directional and axial data', *J. Multivariate Anal.* **60**, 154–175.

Bourguin, S., Durastanti, C., Marinucci, D. & Peccati, G. (2016), 'Gaussian approximation of nonlinear statistics on the sphere', *J. Math. Anal. Appl.* **436**, 1121–1148.

Breitenberger, E. (1963), 'Analogues of the normal distribution on the circle and the sphere', *Biometrika* **50**, 81–88.

Cai, T., Fan, J. & Jiang, T. (2013), 'Distributions of angles in random packing on spheres', *J. Mach. Learn. Res.* **14**, 1837–1864.

Cai, T. & Jiang, T. (2012), 'Phase transition in limiting distributions of coherence of high-dimensional random matrices', *J. Multivariate Anal.* **107**, 24–39.

Cai, T., Ma, Z. & Wu, Y. (2013), 'Sparse PCA: Optimal rates and adaptive estimation', *Ann. Statist.* **41**, 3074–3110.

Cannings, T. I. & Samworth, R. J. (2017), Random projection ensemble classification (with discussion). *J. Roy. Stat. Soc. Ser. B*, forthcoming.

Carnicero, J. A., Ausín, M. C. & Wiper, M. P. (2013), 'Non-parametric copulas for circular–linear and circular–circular data: an application to wind directions', *Stoch. Environ. Res. Risk Assess.* **27**, 1991–2002.

Cartwright, D. E. (1963), The use of directional spectra in studying the output of a wave recorder on a moving ship, *in* 'Ocean Wave Spectra', Englewood Cliffs, NJ: Prentice-Hall, pp. 203–218.

Chang, T. (2004), 'Spatial statistics', *Statist. Sci.* **19**, 624–635.

Chang, T. & Rivest, L.-P. (2001), 'M-estimation for location and regression parameters in group models: a case study using Stiefel manifolds', *Ann. Statist.* **29**, 784–814.

Chang, T. & Tsai, M.-T. (2003), 'Asymptotic relative Pitman efficiency in group models', *J. Multivariate Anal.* **85**, 395–415.

Chaudhuri, P. & Marron, J. S. (1999), 'SiZer for exploration of structures in curves', *J. Amer. Statist. Assoc.* **94**, 807–823.

Chaudhuri, P. & Marron, J. S. (2000), 'Scale space view of curve estimation', *Ann. Statist.* **28**, 408–428.

Chen, A. & Bickel, P. J. (2006), 'Efficient independent component analysis', *Ann. Statist.* **34**, 2825–2855.

Chen, S. & Qin, Y. (2010), 'A two-sample test for high-dimensional data with applications to gene-set testing', *Ann. Statist.* **38**, 808–835.

Chen, S. X., Zhang, L.-X. & Zhong, P.-S. (2010), 'Tests for high-dimensional covariance matrices', *J. Amer. Statist. Assoc.* **105**, 810–819.

Chen, X., Fan, Y. & Tsyrennikov, V. (2006), 'Efficient estimation of semiparametric multivariate copula models', *J. Amer. Statist. Assoc.* **101**, 1228–1240.

Chikuse, Y. (2012), *Statistics on Special Manifolds*, Vol. 174, Springer Science & Business Media.

Chiu, S.-T. (1996), 'A comparative review of bandwidth selection for kernel density estimation', *Statist. Sci.* **6**, 129–145.

Choi, S., Hall, W. & Schick, A. (1996), 'Asymptotically uniformly most powerful tests in parametric and semiparametric models', *Ann. Statist.* **24**, 841–861.

Christie, D. (2015), 'Efficient von Mises-Fisher concentration parameter estimation using Taylor series', *J. Stat. Comput. Sim.* **85**, 3259–3265.

Cordeiro, G. M. & Ferrari, S. (1991), 'A modified score test statistic having chi-squared distribution to order n^{-1}', *Biometrika* **78**, 573–582.

Croux, C., Filzmoser, P. & Fritz, H. (2013), 'Robust sparse principal component analysis', *Technometrics* **55**, 202–214.

Cuesta-Albertos, J. A., Cuevas, A. & Fraiman, R. (2009), 'On projection-based tests for spherical and compositional data', *Stat. Comput.* **19**, 367–380.

Cutland, N. & Ng, S.-A. (1993), 'The Wiener sphere and Wiener measure', *Ann. Probab.* **21**, 1–13.

Cutting, C., Paindaveine, D. & Verdebout, T. (2017*a*), 'Testing uniformity on high-dimensional spheres against contiguous rotationally symmetric alternatives'. *Ann. Statist.*, forthcoming.

Cutting, C., Paindaveine, D. & Verdebout, T. (2017*b*), Tests of concentration for low-dimensional and high-dimensional directional data, *in* S. Ejaz Ahmed, ed., 'Big and Complex Data Analysis: Methodology and Applications', Springer, New York, forthcoming.

de Leeuw, J., Hornik, K. & Mair, P. (2009), 'Isotone optimization in R: Pool-Adjacent-Violators Algorithm (PAVA) and active set methods', *J. Stat. Softw.* **32**, 1–24.

Deschepper, E., Thas, O. & Ottoy, J. P. (2008), 'Tests and diagnostic plots for detecting lack-of-fit for circular-linear regression models', *Biometrics* **64**, 912–920.

Dhillon, I. & Sra, S. (2003), Modeling data using directional distributions, Technical Report TR-03-06, Department of Computer Sciences, University of Texas at Austin.

Di Marzio, M., Panzera, A. & Taylor, C. C. (2009), 'Local polynomial regression for circular predictors', *Stat. Probab. Lett.* **79**, 2066–2075.

Di Marzio, M., Panzera, A. & Taylor, C. C. (2011), 'Kernel density estimation on the torus', *J. Statist. Plann. Infer.* **141**, 2156–2173.

Di Marzio, M., Panzera, A. & Taylor, C. C. (2012), 'Smooth estimation of circular cumulative distribution functions and quantiles', *J. Nonparam. Statist.* **24**, 935–949.

Di Marzio, M., Panzera, A. & Taylor, C. C. (2013), 'Non-parametric regression for circular responses', *Scand. J. Statist.* **40**, 238–255.

Di Marzio, M., Panzera, A. & Taylor, C. C. (2014), 'Nonparametric regression for spherical data', *J. Amer. Statist. Assoc.* **109**, 748–763.

Di Marzio, M., Panzera, A. & Taylor, C. C. (2016), 'Nonparametric circular quantile regression', *J. Statist. Plann. Infer.* **170**, 1–14.

Diggle, P. J., Fisher, N. I. & Lee, A. J. (1985), 'A comparison of tests of uniformity for spherical data', *Austral. J. Statist.* **27**, 53–59.

Dimroth, E. (1962), 'Untersuchungen zum Mechanismus von Blastesis und syntexis in Phylitten und Hornfelsen des südwestlichen Fichtelgebirges I. Die statistische Auswertung einfacher Gürteldiagramme', *Tscherm. Min. Petr. Mitt.* **8**, 248–274.

Dimroth, E. (1963), 'Fortschritte der Gefügestatistik', *N. Jb. Min. Mh.* **13**, 186–192.

Dominicy, Y. & Veredas, D. (2013), 'The Method of Simulated Quantiles', *J. Economet.* **172**, 208–221.

Downs, T. D. & Mardia, K. V. (2002), 'Circular regression', *Biometrika* **89**, 683–697.

Dryden, I. L. (2005), 'Statistical analysis on high-dimensional spheres and shape spaces', *Ann. Statist.* **33**, 1643–1665.

Dryden, I. L. & Kent, J. T., eds (2015), *Geometry Driven Statistics*, Wiley Series in Probability and Statistics.

Duerinckx, M. & Ley, C. (2012), 'Maximum likelihood characterization of rotationally symmetric distributions on the sphere', *Sankhyā A* **74**, 249–262.

Dyckerhoff, R., Ley, C. & Paindaveine, D. (2015), 'Depth-based runs tests for bivariate central symmetry', *Ann. Inst. Statist. Math.* **67**, 917–941.

Einmahl, J. H. J. & Gan, Z. (2016), 'Testing for central symmetry', *J. Statist. Plann. Infer.* **169**, 27–33.

Fang, K. T., Kotz, S. & Ng, K. W. (1990), *Symmetric Multivariate and Related Distributions*, Chapman & Hall.

Fernández-Durán, J. J. (2004), 'Circular distributions based on nonnegative trigonometric sums', *Biometrics* **60**, 499–503.

Fernández-Durán, J. J. (2007), 'Models for circular-linear and circular-circular data constructed from circular distributions based on nonnegative trigonometric sums', *Biometrics* **63**, 579–585.

Fernández-Durán, J. J. & Gregorio-Domínguez, M. M. (2010), 'Maximum likelihood estimation of nonnegative trigonometric sum models using a Newton-like algorithm on manifolds', *Electron. J. Stat.* **4**, 1402–1410.

Fernández, M., Rueda, C. & Peddada, S. (2012), 'Identification of a core set of signature cell cycle genes whose relative order of time to peak expression is conserved across species', *Nucleic Acids Res.* **40**, 2823–2832.

Ferreira, J. T. A. S., Juárez, M. A. & Steel, M. F. J. (2008), 'Directional log-spline distributions', *Bayesian Anal.* **3**, 297–316.

Figueiredo, A. (2007), 'Comparison of tests of uniformity defined on the hypersphere', *Stat. Probab. Lett.* **77**, 329–334.

Figueiredo, A. & Gomes, P. (2003), 'Power of tests of uniformity defined on the hypersphere', *Comm. Statist. Simul. Comput.* **32**, 87–94.

Fisher, N. I. (1985), 'Spherical medians', *J. Roy. Stat. Soc. Ser. B* **47**, 342–348.

Fisher, N. I. (1993), *Statistical Analysis of Circular Data*, Cambridge University Press.

Fisher, N. I., Lewis, T. & Embleton, B. J. J. (1987), *Statistical Analysis of Spherical Data*, Cambridge University Press, Cambridge, 1st paperback edition (with corrections) (1993).

Fisher, R. A. (1953), 'Dispersion on a sphere', *Proc. R. Soc. Lond. Ser. A* **217**, 295–305.

Francq, C. & Zakoïan, J.-M. (2004), 'Maximum likelihood estimation of pure GARCH and ARMA-GARCH processes', *Bernoulli* **10**, 605–637.

Francq, C. & Zakoïan, J.-M. (2012), 'Strict stationarity testing and estimation of explosive and stationary generalized autoregressive conditional heteroscedasticity models', *Econometrica* **80**, 821–861.

García-Portugués, E. (2013), 'Exact risk improvement of bandwidth selectors for kernel density estimation with directional data', *Electron. J. Stat.* **7**, 1655–1685.

García-Portugués, E., Barros, A. M. G., Crujeiras, R. M., González-Manteiga, W. & Pereira, J. M. C. (2014), 'A test for directional-linear independence, with applications to wildfire orientation and size', *Stoch. Environ. Res. Risk Assess.* **28**, 1261–1275.

García-Portugués, E., Crujeiras, R. M. & González-Manteiga, W. (2013*a*), 'Exploring wind direction and SO2 concentration by circular-linear density estimation', *Stoch. Environ. Res. Risk Assess.* **27**, 1055–1067.

García-Portugués, E., Crujeiras, R. M. & González-Manteiga, W. (2013*b*), 'Kernel density estimation for directional-linear data', *J. Multivariate Anal.* **121**, 152–175.

García-Portugués, E., Crujeiras, R. M. & González-Manteiga, W. (2015), 'Central limit theorems for directional and linear data with applications', *Statist. Sinica* **25**, 1207–1229.

García-Portugués, E., Van Keilegom, I., Crujeiras, R. M. & González-Manteiga, W. (2017), 'Testing parametric models in linear-directional regression'. *Scand. J. Statist.*, forthcoming.

Garel, B. & Hallin, M. (1995), 'Local asymptotic normality of multivariate ARMA processes with a linear trend', *Ann. Inst. Statist. Math.* **47**, 551–579.

Gatto, R. (2008), 'Some computational aspects of the generalized von Mises distribution', *Stat. Comput.* **18**, 321–331.

Gatto, R. (2009), 'Information theoretic results for circular distributions', *Statistics* **43**, 409–421.

Gatto, R. & Jammalamadaka, S. R. (2003), 'Inference for wrapped symmetric alpha-stable circular models', *Sankhyā* **65**, 333–355.

Gatto, R. & Jammalamadaka, S. R. (2007), 'The generalized von Mises distribution', *Statist. Methodol.* **4**, 341–353.

Gauss, C. F. (1809), *Theoria motus corporum coelestium in sectionibus conicis solem ambientium*, Cambridge Library Collection. Cambridge University Press, Cambridge. Reprint of the 1809 original.

Genest, C. & Werker, B. J. (2002), Conditions for the asymptotic semiparametric efficiency of an omnibus estimator of dependence parameters in copula models, *in* C. M. Cuadras, J. Fortiana & J. A. Rodriguez-Lallena, eds, 'Distributions With Given Marginals and Statistical Modelling', Springer, pp. 103–112.

Gijbels, I. & Mielniczuk, J. (1990), 'Estimating the density of a copula function', *Comm. Statist. Theor. Meth.* **19**, 445–464.

Gill, J. & Hangartner, D. (2010), 'Circular data in political science and how to handle it', *Polit. Anal.* **18**, 316–336.

Giné, M. E. (1975), 'Invariant tests for uniformity on compact Riemannian manifolds based on Sobolev norms', *Ann. Statist.* **3**, 1243–1266.

Gradshteyn, I. S. & Ryzhik, I. M. (2015), *Tables of Integrals, Series and Products, 8th Ed.*, Academic Press, London.

Graham, J. W. (1949), 'The stability and significance of magnetism in sedimentary rocks', *J. Geophys. Res.* **54**, 131–167.

Gumbel, E. J., Greenwood, J. A. & Durand, D. (1953), 'The circular normal distribution: theory and tables', *J. Amer. Statist. Assoc.* **48**, 131–152.

Guo, Y., Hastie, T. & Tibshirani, R. (2007), 'Regularized linear discriminant analysis and its application in microarrays', *Biostatistics* **8**, 86–100.

Hall, P., Watson, G. S. & Cabrera, J. (1987), 'Kernel density estimation with spherical data', *Biometrika* **74**, 751–762.

Hallin, M. & Ley, C. (2012*a*), Permutation tests, *in* 'Encyclopedia of Environmetrics Second Edition, A.-H. El-Shaarawi and W. Piegorsch (eds)', John Wiley & Sons Ltd, Chichester, UK, pp. 209–210.

Hallin, M. & Ley, C. (2012*b*), 'Skew-symmetric distributions and Fisher information — a tale of two densities', *Bernoulli* **18**, 747–763.

Hallin, M. & Paindaveine, D. (2002), 'Optimal tests for multivariate location based on interdirections and pseudo-Mahalanobis ranks', *Ann. Statist.* **30**, 1103–1133.

Hallin, M. & Paindaveine, D. (2004), 'Rank-based optimal tests of the adequacy of an elliptic VARMA model', *Ann. Statist.* **32**, 2642–2678.

Hallin, M. & Paindaveine, D. (2006), 'Semiparametrically efficient rank-based inference for shape. I. Optimal rank-based tests for sphericity', *Ann. Statist.* **34**, 2707–2756.

Hallin, M. & Paindaveine, D. (2008), 'A general method for constructing pseudo-Gaussian tests', *J. Japan Statist. Soc.* **38**, 27–39.

Hallin, M., Paindaveine, D. & Verdebout, T. (2010), 'Optimal rank-based testing for principal components', *Ann. Statist.* **38**, 3245–3299.

Hallin, M., Taniguchi, M., Serroukh, A. & Choy, K. (1999), 'Local asymptotic normality for regression models with long-memory disturbance', *Ann. Statist.* **27**, 2054–2080.

Hallin, M., Van Den Akker, R. & Werker, B. J. (2011), 'A class of simple distribution-free rank-based unit root tests', *J. Economet.* **163**, 200–214.

Hallin, M. & Werker, B. J. (2003), 'Semi-parametric efficiency, distribution-freeness and invariance', *Bernoulli* **9**, 137–165.

Hamelryck, T., Mardia, K. V. & Ferkinghoff-Borg, J., eds (2012), *Bayesian Methods in Structural Bioinformatics*, Springer-Verlag Berlin Heidelberg.

Han, F. & Liu, H. (2014), 'Scale-invariant sparse PCA on high-dimensional meta-elliptical data', *J. Amer. Statist. Assoc.* **109**, 275–287.

Harder, T., Boomsma, W., Paluszewski, M., Frellsen, J., Johansson, K. E. & Hamelryck, T. (2010), 'Beyond rotamers: a generative, probabilistic model of side chains in proteins', *BMC Bioinformatics* **11**, 306.

Hayakawa, T. (1990), 'On tests for the mean direction of the Langevin distribution', *Ann. Inst. Statist. Math.* **42**, 359–373.

Hayakawa, T. & Puri, M. L. (1985), 'Asymptotic expansions of the distributions of some test statistics', *Ann. Inst. Statist. Math.* **37**, 95–108.

Hornik, K. & Grün, B. (2014), 'On maximum likelihood estimation of the concentration parameter of von Mises–Fisher distributions', *Comput. Stat.* **29**, 945–957.

Huckemann, S., Kim, K.-R., Munk, A., Rehfeldt, F., Sommerfeld, M., Weickert, J. & Wollnik, C. (2016), 'The circular SiZer, inferred persistence of shape parameters and application to stem cell stress fibre structures', *Bernoulli* **22**, 2113–2142.

Isham, V. (1977), 'A Markov construction for a multidimensional point process', *J. Appl. Probab.* **14**, 507–515.

Jammalamadaka, S. R. & Kozubowski, T. J. (2003), 'A new family of circular models: the wrapped Laplace distributions', *Adv. Appl. Stat.* **3**, 77–103.

Jammalamadaka, S. R. & Kozubowski, T. J. (2004), 'New families of wrapped distributions for modeling skew circular data', *Comm. Statist. Theor. Meth.* **33**, 2059–2074.

Jammalamadaka, S. R. & SenGupta, A. (2001), *Topics in Circular Statistics*, World Scientific, Singapore.

Jeffreys, H. (1948), *Theory of Probability, 2nd ed*, Oxford University Press, Oxford.

Jiang, T. & Yang, F. (2013), 'Central limit theorems for classical likelihood ratio tests for high-dimensional normal distributions', *Ann. Statist.* **41**, 1693–2262.

Joe, H. (1997), *Multivariate Models and Dependence Concepts*, Chapman & Hall, London.

Johnson, R. A. & Wehrly, T. E. (1977), 'Measures and models for angular correlation and angular-linear correlation', *J. Roy. Stat. Soc. Ser. B* **39**, 222–229.

Johnson, R. A. & Wehrly, T. E. (1978), 'Some angular-linear distributions and related regression models', *J. Amer. Statist. Assoc.* **73**, 602–606.

Johnstone, I. M. (2001), 'On the distribution of the largest eigenvalue in principal components analysis', *Ann. Statist.* **29**, 295–327.

Jones, M. C. (2015), 'On families of distributions with shape parameters (with discussion)', *Internat. Statist. Rev.* **83**, 175–192.

Jones, M. C. & Anaya-Izquierdo, K. (2011), 'On parameter orthogonality in symmetric and skew models', *J. Statist. Plann. Infer.* **141**, 758–770.

Jones, M. C. & Pewsey, A. (2005), 'A family of symmetric distributions on the circle', *J. Amer. Statist. Assoc.* **100**, 1422–1428.

Jones, M. C. & Pewsey, A. (2012), 'Inverse Batschelet distributions for circular data', *Biometrics* **68**, 183–193.

Jones, M. C., Pewsey, A. & Kato, S. (2015), 'On a class of circulas: copulas for circular distributions', *Ann. Inst. Statist. Math.* **67**, 843–862.

Jung, S., Dryden, I. L. & Marron, J. S. (2012), 'Analysis of principal nested spheres', *Biometrika* **99**, 551–568.

Jupp, P. E. (2001), 'Modifications of the Rayleigh and Bingham tests for uniformity of directions', *J. Multivariate Anal.* **77**, 1–20.

Jupp, P. E. (2008), 'Data-driven Sobolev tests of uniformity on compact Riemannian manifolds', *Ann. Statist.* **36**, 1246–1260.

Jupp, P. E. (2009), 'Data-driven tests of uniformity on product manifolds', *J. Statist. Plann. Infer.* **139**, 3820–3829.

Jupp, P. E. (2015), 'Copulae on products of compact Riemannian manifolds', *J. Multivariate Anal.* **140**, 92–98.

Jupp, P. E. & Mardia, K. V. (1989), 'A unified view of the theory of directional statistics, 1975-1988', *Internat. Statist. Rev.* **57**, 261–294.

Jupp, P. E. & Spurr, B. D. (1983), 'Sobolev tests for symmetry of directional data', *Ann. Statist.* **11**, 1225–1231.

Kato, S. (2009), 'A distribution for a pair of unit vectors generated by Brownian motion', *Bernoulli* **15**, 898–921.

Kato, S. & Jones, M. C. (2010), 'A family of distributions on the circle with links to, and applications arising from, Möbius transformation', *J. Amer. Statist. Assoc.* **105**, 249–262.

Kato, S. & Jones, M. C. (2013), 'An extended family of circular distributions related to wrapped Cauchy distributions via Brownian motion', *Bernoulli* **19**, 154–171.

Kato, S. & Jones, M. C. (2015), 'A tractable and interpretable four-parameter family of unimodal distributions on the circle', *Biometrika* **102**, 181–190.

Kato, S. & Pewsey, A. (2015), 'A Möbius transformation-induced distribution on the torus', *Biometrika* **102**, 359–370.

Kato, S. & Shimizu, K. (2008), 'Dependent models for observations which include angular ones', *J. Statist. Plann. Infer.* **138**, 3538–3549.

Kato, S., Shimizu, K. & Shieh, G. S. (2008), 'A circular-circular regression model', *Statist. Sinica* **18**, 633–645.

Kent, J. T. (1982), 'The Fisher-Bingham distribution on the sphere', *J. Roy. Stat. Soc. Ser. B* **44**, 71–80.

Kent, J. T., Mardia, K. V. & Taylor, C. C. (2008), Modelling strategies for bivariate circular data, *in* S. Barber, P. D. Baxter, A. Gusnanto & K. V. Mardia, eds, 'LASR 2008: Art Sci. Statist. Bioinformatics', pp. 70–73.

Kim, P. T., Koo, J.-Y. & Pham Ngoc, T. M. (2016), 'Supersmooth testing on the sphere over analytic classes', *J. Nonparam. Statist.* **28**, 84–115.

Klemelä, J. (2000), 'Estimation of densities and derivatives of densities with directional data', *J. Multivariate Anal.* **73**, 18–40.

Klugkist, I. G., Bullens, J. & Postma, A. (2012), 'Evaluating order constrained hypotheses for circular data using permutation tests', *Brit. J. Math. Stat. Psy.* **65**, 222–236.

Koenker, R. (2005), *Quantile Regression, 1st edition*, Cambridge University Press.

Koul, H. L. & Schick, A. (1997), 'Efficient estimation in nonlinear autoregressive time-series models', *Bernoulli* **3**, 247–277.

Kuiper, N. H. (1960), 'Tests concerning random points on a circle', *Ned. Akad. Wet. Proc. Ser. A* **63**, 38–47.

Kume, A. & Wood, A. T. A. (2005), 'Saddlepoint approximations for the Bingham and Fisher-Bingham normalising constants', *Biometrika* **92**, 465–476.

Lacour, C. & Pham Ngoc, T. M. (2014), 'Goodness-of-fit test for noisy directional data', *Bernoulli* **20**, 2131–2168.

Langevin, P. (1905*a*), 'Magnétisme et théorie des électrons', *Ann. Chim. Phys.* **5**, 70–127.

Langevin, P. (1905*b*), 'Sur la théorie du magnétisme', *J. Phys.* **4**, 678–693.

LaRiccia, V. N. (1991), 'Smooth goodness-of-fit tests: a quantile function approach', *J. Amer. Statist. Assoc.* **86**, 427–431.

Le Cam, L. (1960), 'Locally asymptotically normal families of distributions', *University of California Publications in Statistics* **3**, 37–98.

Le Cam, L. & Yang, G. L. (2000), *Asymptotics in Statistics: Some Basic Concepts*, Springer Series in Statistics, New York.

Ledoit, O. & Wolf, M. (2002), 'Some hypothesis tests for the covariance matrix when the dimension is large compared to the sample size', *Ann. Statist.* **30**, 1081–1102.

Lévy, P. (1939), 'L'addition des variables aléatoires définies sur une circonférence', *Bull. Soc. Math. France* **67**, 1–41.

Lewis, T. & Fisher, N. I. (1982), 'Graphical methods for investigating the fit of a Fisher distribution to spherical data', *Geophys. J. R. Astr. Soc.* **69**, 1–13.

Ley, C. (2015), 'Flexible modelling in statistics: past, present and future', *Journal de la Société Française de Statistique* **156**, 76–96.

Ley, C. & Paindaveine, D. (2010*a*), 'Multivariate skewing mechanisms: a unified perspective based on the transformation approach', *Stat. Probab. Lett.* **80**, 1685–1694.

Ley, C. & Paindaveine, D. (2010*b*), 'On the singularity of multivariate skew-symmetric models', *J. Multivariate Anal.* **101**, 1434–1444.

Ley, C., Paindaveine, D. & Verdebout, T. (2015), 'High-dimensional tests for spherical location and spiked covariance', *J. Multivariate Anal.* **139**, 79–91.

Ley, C., Sabbah, C. & Verdebout, T. (2014), 'A new concept of quantiles for directional data and the angular Mahalanobis depth', *Electron. J. Stat.* **8**, 795–816.

Ley, C., Swan, Y., Thiam, B. & Verdebout, T. (2013), 'Optimal R-estimation of a spherical location', *Statist. Sinica* **23**, 305–332.

Ley, C., Swan, Y. & Verdebout, T. (2017), 'Efficient ANOVA for directional data', *Ann. Inst. Statist. Math.* **69**, 39–62.

Ley, C. & Verdebout, T. (2014*a*), 'Local powers of optimal one- and multi-sample tests for the concentration of Fisher–von Mises–Langevin distributions', *Internat. Statist. Rev.* **82**, 440–456.

Ley, C. & Verdebout, T. (2014*b*), 'Simple optimal tests for circular reflective symmetry about a specified median direction', *Statist. Sinica* **14**, 1319–1340.

Ley, C. & Verdebout, T. (2017), 'Skew-rotationally-symmetric distributions and related efficient inferential procedures'. *J. Multivariate Anal.*, forthcoming.

Li, J. & Chen, S. X. (2012), 'Two-sample tests for high-dimensional covariance matrices', *Ann. Statist.* **40**, 908–940.

Liu, R. Y. (1990), 'On a notion of data depth based on random simplices', *Ann. Statist.* **18**, 405–414.

Liu, R. Y. & Singh, K. (1992), 'Ordering directional data: concept of data depth on circles and spheres', *Ann. Statist.* **20**, 1468–1484.

Maksimov, V. M. (1967), 'Necessary and sufficient statistics for a family of shifts of probability distributions on continuous bicompact groups', *Theory Probab. Appl.* **12**, 267–280.

Mardia, K. V. (1972), *Statistics of Directional Data*, Academic Press, New York.

Mardia, K. V. (1975), 'Statistics of directional data (with discussion)', *J. Roy. Stat. Soc. Ser. B* **37**, 349–393.

Mardia, K. V., Hughes, G., Taylor, C. C. & Singh, H. (2008), 'A multivariate von Mises distribution with applications to bioinformatics', *Canad. J. Statist.* **36**, 99–109.

Mardia, K. V. & Jupp, P. E. (2000), *Directional Statistics*, Wiley, New York.

Mardia, K. V. & Patrangenaru, V. (2005), 'Directions and projective shapes', *Ann. Statist.* **33**, 1666–1699.

Mardia, K. V. & Sutton, T. W. (1975), 'On the modes of a mixture of two von Mises distributions', *Biometrika* **62**, 699–701.

Mardia, K. V. & Sutton, T. W. (1978), 'A model for cylindrical variables with applications', *J. Roy. Stat. Soc. Ser. B* **40**, 229–233.

Mardia, K. V., Taylor, C. C. & Subramaniam, G. K. (2007), 'Protein bioinformatics and mixtures of bivariate von Mises distributions for angular data', *Biometrics* **63**, 505–512.

Marinucci, D. & Peccati, G. (2011), *Random Fields on the Sphere: Representation, Limit Theorems and Cosmological Applications*, Vol. 389, London Mathematical Society Lecture Notes Series, Cambridge University Press.

McLachlan, G. J. & Peel, D. (2000), *Finite Mixture Models*, Wiley Series in Probability and Statistics.

McWilliams, T. P. (1990), 'A distribution-free test for symmetry based on a runs statistic', *J. Amer. Statist. Assoc.* **85**, 1130–1133.

Minh, D. L. P. & Farnum, N. R. (2003), 'Using bilinear transformations to induce probability distributions', *Comm. Statist. Theor. Meth.* **32**, 1–9.

Modarres, R. & Gastwirth, J. L. (1996), 'A modified runs test for symmetry', *Stat. Probab. Lett.* **31**, 107–112.

Mooney, J. A., Helms, P. J. & Jolliffe, I. T. (2003), 'Fitting mixtures of von Mises distributions: a case study involving sudden infant death syndrome', *Comp. Stat. Data Anal.* **41**, 505–513.

Neeman, T. & Chang, T. (2001), 'Rank score statistics for spherical data', *Contem. Math.* **287**, 241–254.

Nelsen, R. B. (2010), *An Introduction to Copulas, 2nd edition*, Springer, New York.

Oliveira, M., Crujeiras, R. M. & Rodríguez-Casal, A. (2012), 'A plug-in rule for bandwidth selection in circular density estimation', *Comp. Stat. Data Anal.* **56**, 3898–3908.

Oliveira, M., Crujeiras, R. M. & Rodríguez-Casal, A. (2014*a*), 'CircSiZer: an exploratory tool for circular data', *Environ. Ecol. Stat.* **21**, 143–159.

Oliveira, M., Crujeiras, R. M. & Rodríguez-Casal, A. (2014*b*), 'NPCirc: an R package for nonparametric circular methods', *J. Stat. Softw.* **61**, 1–26.

Onatski, A., Moreira, M. & Hallin, M. (2013), 'Asymptotic power of sphericity tests for high-dimensional data', *Ann. Statist.* **41**, 1204–1231.

Onatski, A., Moreira, M. J. & Hallin, M. (2014), 'Signal detection in high dimension: The multispiked case', *Ann. Statist.* **42**, 225–254.

Oualkacha, K. & Rivest, L.-P. (2009), 'A new statistical model for random unit vectors', *J. Multivariate Anal.* **100**, 70–80.

Paindaveine, D. (2012), Elliptical symmetry, *in* A. H. El-Shaarawi & W. Piegorsch, eds, 'Encyclopedia of Environmetrics, 2nd edition', John Wiley and Sons Ltd, Chichester, UK, pp. 802–807.

Paindaveine, D. & Van Bever, G. (2015), 'Nonparametrically consistent depth-based classifiers', *Bernoulli* **21**, 62–82.

Paindaveine, D. & Verdebout, T. (2015), Optimal rank-based tests for the location parameter of a rotationally symmetric distribution on the hypersphere, *in* M. Hallin, D. Mason, D. Pfeifer & J. Steinebach, eds, 'Mathematical Statistics and Limit Theorems: Festschrift in Honor of Paul Deheuvels', Springer, pp. 249–269.

Paindaveine, D. & Verdebout, T. (2016), 'On high-dimensional sign tests', *Bernoulli* **22**, 1745–1769.

Paindaveine, D. & Verdebout, T. (2017), 'Inference on the mode of weak directional signals: a Le Cam perspective on hypothesis testing near singularities'. *Ann. Statist.*, forthcoming.

Papakonstantinou, V. (1979), Beiträge zur zirkulären Statistik, PhD thesis, University of Zurich.

Partlett, C. & Patil, P. (2017), 'Measuring asymmetry and testing symmetry'. *Ann. Inst. Statist. Math.*, forthcoming.

Parzen, E. (1962), 'On estimation of a probability density function and mode', *Ann. Math. Statist.* **33**, 1065–1076.

Peel, D., Whiten, W. J. & McLachlan, G. J. (2001), 'Finite mixtures of Kent distributions to aid in joint set identification', *J. Amer. Statist. Assoc.* **96**, 56–63.

Pewsey, A. (2000), 'The wrapped skew-normal distribution on the circle', *Comm. Statist. Theor. Meth.* **29**, 2459–2472.

Pewsey, A. (2002), 'Testing circular symmetry', *Canad. J. Statist.* **30**, 591–600.

Pewsey, A. (2004), 'Testing for circular reflective symmetry about a known median axis', *J. Appl. Stat.* **31**, 575–585.

Pewsey, A. (2008), 'The wrapped stable family of distributions as a flexible model for circular data', *Comp. Stat. Data Anal.* **52**, 1516–1523.

Pewsey, A. & Kato, S. (2016), 'Parametric bootstrap goodness-of-fit testing for Wehrly–Johnson bivariate circular distributions', *Stat. Comput.* **26**, 1307–1317.

Pewsey, A., Lewis, T. & Jones, M. C. (2007), 'The wrapped t family of circular distributions', *Aust. NZ. J. Stat.* **49**, 79–91.

Pewsey, A., Neuhäuser, M. & Ruxton, G. D. (2013), *Circular Statistics in R*, Oxford University Press, Oxford.

Pewsey, A., Shimizu, K. & de la Cruz, R. (2011), 'On an extension of the von Mises distribution due to Batschelet', *J. Appl. Stat.* **38**, 1073–1085.

Presnell, B., Morrison, S. P. & Littell, R. C. (1998), 'Projected multivariate linear models for directional data', *J. Amer. Statist. Assoc.* **93**, 1068–1077.

Purkayastha, S. (1991), 'A rotationally symmetric directional distribution obtained through maximum likelihood characterization', *Sankhyā A* **53**, 70–83.

Pycke, J.-R. (2007), 'A decomposition for invariant tests of uniformity on the sphere', *Proc. Amer. Math. Soc.* **135**, 2983–2993.

Ramachandran, G. N., Ramakrishnan, C. & Sasisekharan, V. (1963), 'Stereochemistry of polypeptide chain configurations', *Molec. Biol.* **7**, 95–99.

Randles, R. H., Fligner, M. A., Policello, G. E. & Wolfe, D. A. (1980), 'An asymptotically distribution-free test for symmetry versus asymmetry', *J. Amer. Statist. Assoc.* **75**, 168–172.

Rayleigh, L. (1880), 'On the resultant of a large number of vibrations of the same pitch and of arbitrary phase', *Phil. Mag.* **10**, 73–78.

Rayleigh, L. (1919), 'On the problem of random vibrations and random flights in one, two and three dimensions', *Phil. Mag.* **37**, 321–347.

Reed, W. & Pewsey, A. (2009), 'Two nested families of skew-symmetric circular distributions', *Test* **18**, 516–528.

Rivest, L.-P. (1988), 'A distribution for dependent unit vectors', *Comm. Statist. Theor. Meth.* **17**, 461–483.

Robertson, T., Wright, F. T. & Dykstra, R. L. (1988), *Order Restricted Statistical Inference*, Wiley, New York.

Rosenblatt, M. (1956), 'Remarks on some nonparametric estimates of a density function', *Ann. Math. Statist.* **27**, 832–837.

Rothman, E. D. & Woodrofe, M. (1972), 'A Cramér-von Mises type statistic for testing symmetry', *Ann. Math. Statist.* **43**, 2035–2038.

Rueda, C., Fernández, M., Barragán, S. & Peddada, S. (2015), Some advances in constrained inference for ordered circular parameters in oscillatory systems, *in* I. L. Dryden & J. T. Kent, eds, 'Geometry Driven Statistics', Wiley Series in Probability and Statistics, pp. 97–114.

Rueda, C., Fernández, M. & Peddada, S. (2009), 'Estimation of parameters subject to order restrictions on a circle with application to estimation of phase angles of cell-cycle genes', *J. Amer. Statist. Assoc.* **104**, 338–347.

Rukhin, A. L. (1972), 'Some statistical decisions about distributions on a circle for large samples', *Sankhyā* **34**, 243–250.

Schach, S. (1969), 'Nonparametric symmetry tests for circular distributions', *Biometrika* **56**, 571–577.

Scornet, E., Biau, G. & Vert, J.-P. (2015), 'Consistency of random forests', *Ann. Statist.* **43**, 1716–1741.

Scott, D. W. (2015), *Multivariate Density Estimation: Theory, Practice, and Visualization, 2nd Edition*, John Wiley & Sons, Inc.

Segers, J., Akker, R. v. d. & Werker, B. J. (2014), 'Semiparametric Gaussian copula models: Geometry and rank-based efficient estimation', *Ann. Statist.* **42**, 1911–1940.

Selby, B. (1964), 'Girdle distributions on a sphere', *Biometrika* **51**, 381–392.

SenGupta, A. & Pal, C. (2001), 'On optimal tests for isotropy against the symmetric wrapped stable-circular uniform mixture family', *J. Appl. Stat.* **28**, 129–143.

Shieh, G. S. & Johnson, R. A. (2005), 'Inference based on a bivariate distribution with von Mises marginals', *Ann. Inst. Statist. Math.* **57**, 789–802.

Shieh, G. S., Zheng, S., Johnson, R. A., Chang, Y. F., Shimizu, K., Wang, C. C. & Tang, S. L. (2011), 'Modeling and comparing the organization of circular genomes', *Bioinformatics* **27**, 912–918.

Shimizu, K. & Iida, K. (2002), 'Pearson type VII distributions on spheres', *Comm. Statist. Theor. Meth.* **31**, 513–526.

Siew, H.-Y. & Shimizu, K. (2008), 'The generalized Laplace distribution on the sphere', *Statist. Methodol.* **5**, 487–501.

Siew, H.-Y., Shimizu, K. & Kato, S. (2008), 'Generalized t-distribution on the circle', *Jpn. J. Appl. Stat.* **37**, 1–16.

Silverman, B. W. (1986), *Density Estimation for Statistics and Data Analysis*, Monographs on Statistics and Applied Probability, Chapman & Hall, London.

Singh, H., Hnizdo, V. & Demchuk, E. (2002), 'Probabilistic model for two dependent circular variables', *Biometrika* **89**, 719–723.

Small, C. G. (1987), 'Measures of centrality for multivariate and directional distributions', *Canad. J. Statist.* **15**, 31–39.

Song, H., Liu, J. & Wang, G. (2012), 'High-order parameter approximation for von Mises–Fisher distributions', *Appl. Math. Comput.* **218**, 11880–11890.

Spurr, B. D. & Koutbeiy, M. A. (1991), 'A comparison of various methods of estimating the parameters in mixtures of von Mises distributions', *Comm. Statist. Simul. Comput.* **20**, 725–741.

Sra, S. (2011), 'A short note on parameter approximation for von Mises–Fisher distributions And a fast implementation of $i_s(x)$', *Comput. Stat.* **27**, 177–190.

Sra, S. & Karp, D. (2013), 'The multivariate Watson distribution: maximum-likelihood estimation and other aspects', *J. Multivariate Anal.* **114**, 256–269.

Srivastava, M. S. & Kubokawa, T. (2013), 'Tests for multivariate analysis of variance in high dimension under non-normality', *J. Multivariate Anal.* **115**, 204–216.

Stephens, M. A. (1969), 'A goodness-of-fit statistic for the circle, with some comparisons', *Biometrika* **56**, 161–168.

Su, Y. & Wu, X.-K. (2011), Smooth test for uniformity on the surface of a unit sphere, *in* 'Proceedings of the 2011 International Conference on Machine Learning and Cybernetics, Guilin, 10-13 July', pp. 867–872.

Sugasawa, S., Shimizu, K. & Kato, S. (2015), A flexible family of distributions on the cylinder. arXiv:1501.06332v2.

Taijeron, H. J., Gibson, A. G. & Chandler, C. (1994), 'Spline interpolation and smoothing on hyperspheres', *SIAM J. Sci. Comput.* **15**, 1111–1125.

Tanabe, A., Fukumizu, K., Oba, S., Takenouchi, T. & Ishii, S. (2007), 'Parameter estimation for von Mises-Fisher distributions', *Comput. Stat.* **22**, 145–157.

Taylor, C. C. (2008), 'Automatic bandwidth selection for circular density estimation', *Comp. Stat. Data Anal.* **52**, 3493–3500.

Taylor, C. C., Mardia, K. V., Di Marzio, M. & Panzera, A. (2012), 'Validating protein structure using kernel density estimates', *J. Appl. Stat.* **39**, 2379–2388.

Toyoda, Y., Suga, K., Murakami, K., Hasegawa, H., Shibata, S., Domingo, V., Escobar, I., Kamata, K., Bradt, H., Clark, G. & La Pointe, M. (1965), 'Studies of primary cosmic rays in the energy region 10^{14} eV to 10^{17} eV (Bolivian Air Shower Joint Experiment)', *Proc. Int. Conf. Cosmic Rays (London)* **2**, 708–711.

Tsai, M.-T. & Sen, P. K. (2007), 'Locally best rotation-invariant rank tests for modal location', *J. Multivariate Anal.* **98**, 1160–1179.

Umbach, D. & Jammalamadaka, S. R. (2009), 'Building asymmetry into circular distributions', *Stat. Probab. Lett.* **79**, 659–663.

van der Vaart, A. (1998), *Asymptotic Statistics*, Cambridge University Press.

van der Vaart, A. (2002), 'The statistical work of Lucien Le Cam', *Ann. Statist.* **30**, 631–682.

Verdebout, T. (2015), 'On some validity-robust tests for the homogeneity of concentrations on spheres', *J. Nonparam. Statist.* **27**, 372–383.

von Mises, R. (1918), 'Uber die 'Ganzzahligkeit' der Atomgewichte und verwandte Fragen', *Phys. Z.* **19**, 490–500.

Wallis, K. F. (2014), 'The two-piece normal, binormal, or double Gaussian distribution: its origin and rediscoveries', *Statist. Sci.* **29**, 106–112.

Wand, M. P. & Jones, M. C. (1995), *Kernel Smoothing*, Vol. 60 of Monographs on Statistics and Applied Probability, Chapman & Hall, London.

Wang, M.-Z. & Shimizu, K. (2012), 'On applying Möbius transformation to cardioid random variables', *Statist. Methodol.* **9**, 604–614.

Watamori, Y. & Jupp, P. E. (2005), 'Improved likelihood ratio and score tests on concentration parameters of von Mises-Fisher distributions', *Stat. Probab. Lett.* **72**, 93–102.

Watson, G. S. (1961), 'Goodness-of-fit tests on a circle', *Biometrika* **48**, 109–114.

Watson, G. S. (1965), 'Equatorial distributions on a sphere', *Biometrika* **52**, 193–201.

Watson, G. S. (1983), *Statistics on Spheres*, Wiley, New York.

Wehrly, T. E. & Johnson, R. A. (1980), 'Bivariate models for dependence of angular observations and a related Markov process', *Biometrika* **66**, 255–256.

Witten, D. M. & Tibshirani, R. (2011), 'Penalized classification using Fisher's linear discriminant', *J. Roy. Stat. Soc. Ser. B* **73**, 753–772.

Yfantis, E. A. & Borgman, L. E. (1982), 'An extension of the von Mises distribution', *Comm. Statist. Theor. Meth.* **11**, 1695–1706.

Zhao, L. & Wu, C. (2001), 'Central limit theorem for integrated squared error of kernel estimators of spherical density', *Sci. China Ser. A* **44**, 474–483.

Zuo, Y. & Serfling, R. J. (2000), 'General notions of statistical depth function', *Ann. Statist.* **28**, 461–482.

Index